Soil Micro-organisms

T. R. G. GRAY *and*
S. T. WILLIAMS
Lecturers in Botany, The University of Liverpool

OLIVER & BOYD
EDINBURGH

OLIVER & BOYD
Tweeddale Court, 14 High Street,
Edinburgh EH1 1YL
A Division of Longman Group Limited.

First Published 1971

ISBN 0 05 002322 5

Printed in Great Britain by
T. and A. Constable Ltd., Edinburgh

PREFACE

Soil is one of the earth's most important resources for within it are held the nutrients and water for growth of plants, including those which provide us with materials for food and shelter. At the present time, when the world's population is increasing rapidly, the demands made upon the soil are also increasing rapidly; larger and larger crop yields are both required and being obtained, partly through the use of new varieties of plants, partly through more intensive and extensive cultivation and partly through the application to soil of new and exotic chemicals. Since soil is also the home of many micro-organisms, it is important that the effects of these changes on the microflora be thoroughly understood, for the microflora can affect plant growth directly by causing disease or indirectly through the recycling of nutrients locked up in dead organic materials.

Unfortunately, the study of the soil microflora is complicated because soil is a heterogeneous environment which is difficult to simulate in the laboratory and because the microflora is so varied and capable of carrying out so many chemical transformations. Any book that attempted to deal with all aspects of such a system would be enormous, and in trying to write an introduction to the subject, we are conscious that many interesting and important topics have been omitted. We have tried to stress the activities of the microflora in nature rather than in the laboratory and have included topics such as the colonisation and utilisation of organic substrates, interactions between micro-organisms, the environment and plants.

The book is intended mainly for undergraduate students taking courses in microbiology, soil science or botany, but it is hoped that both school teachers and post-graduate students will find it useful. The stimulus for writing this book came mainly from our students, but many other people have helped us in writing it, including the authors and publishers who have allowed us to reproduce many of the photographs, figures and tables. We wish especially to thank Professor V. H. Heywood, who gave us much encouragement and constructive criticism during all stages of the book's production and Professors E. W. Russell, N. A. Burges, D. H. Jennings and D. Parkinson, who commented on several of the chapters. Much of the book was written while one of us (T. R. G. G.) was on leave at the University of Minnesota and we wish to thank the staff of the Departments of Microbiology and Soil Science in that University for the facilities they provided and, in particular, Dr E. L. Schmidt for many useful discussions. The manuscript was typed entirely by Mrs Rosemary Featherstone and many of the photographs were taken by Mr A. J. Tollitt.

During the last three years we have spent many evenings discussing the book and we would both like to thank Kath Williams for putting up with this so patiently.

CONTENTS

1

The Soil Population

Man is almost entirely dependent upon plants as a source of food for himself and for the growth of animals. Many of these plants are rooted in soil, from which they obtain essential mineral nutrients that originate from dead plants, animal remains and soil and rock minerals. The release of nutrients from these materials is brought about by physical, chemical and biological processes, but it is doubtful whether the rate of release would be adequate to maintain more than small plant populations if no biological activity occurred. Soil is populated by many organisms, including animals and micro-organisms, but it is generally considered that it is the micro-organisms which play the most important role in the release of minerals and carbon dioxide for plant growth. In this and subsequent chapters, we shall consider the diversity of these micro-organisms and the variety of processes in which they participate.

One of the most striking features of the soil microflora is its diversity since fungi, bacteria, actinomycetes, algae and viruses, belonging to innumerable genera and species, can be found in almost any soil sample. The relative proportions of the different groups are influenced to some extent by the environment. Thus fungi are dominant in many acidic forest soils while bacteria predominate in waterlogged muds. Actinomycetes are seldom numerically dominant, although they comprised up to 95 per cent of the microflora in the soils of the Bikini Atoll before and after the explosion of an atomic bomb. Fortunately, these latter observations have never been made upon more familiar soils! Dominance of algae may also occur after catastrophic events, for they were amongst the initial colonisers of newly forming soils, following the volcanic explosions at Krakatao in 1883 and Surtsey in 1965. However, they are found as pioneer colonisers in many other environments, especially if the soil is wet.

The various propagules of these organisms may be present in large quantities in normal soils and even a single gram of soil may contain 5 metres of fungal mycelium, 10^8 bacterial cells and 10^6 spores of actinomycetes, to say nothing of the algae and viruses. Given that the average weight of a bacterial cell or actinomycete spore is $1{\cdot}5 \times 10^{-12}$ g and that 1 metre of fungal mycelium weighs $9{\cdot}4 \times 10^{-5}$ g[6], the total biomass of the microflora will be about $6{\cdot}0 \times 10^{-4}$ g, that is less than 0·06 per cent of the total weight of the soil. Even this is probably an overestimate since figures for the average weight of cells have been determined for organisms grown in the laboratory. Thus, if counts of the various propagules are considered without reference to biomass, a misleading impression of their abundance is obtained. Even in a sandy soil where the surface area of the soil particles is low (*c.* 72 cm^2 per g), only about 0·02 per cent of the particle surfaces are colonised by bacteria, so that the impression gained from observation of such soil particles (Plate I (a)) is not unlike that obtained of the vegetation seen from an aeroplane flying over an arid desert.

The continued existence of this complex microflora is dependent upon the presence of sources of food and energy. In most cases, it is the soil organic matter—the remains of plant leaves, roots and stems, the secretions of living roots and the corpses of dead animals and micro-organisms—which constitute this source. This organic matter is usually dissolved in the soil water or else exists as particles of varying size. Generally, soils contain between 1 and 3 per cent by weight of organic matter. Not all of this is available to the micro-organisms and much of the microflora has to grow and survive on a relatively small amount of food, spread over the extensive surface area of all the soil particles. It is hardly surprising that many microbiologists consider the microflora to be at or near starvation level.

Gause[145] suggested that when different micro-organisms compete with one another for the same environment or niche, one will eventually oust the others, but this does not mean that soil will contain only one or a few types of organisms. The vast array of food materials, the large surfaces presented for colonisation and the continuous fluctuation in environmental conditions result in the development of many distinct ecological niches in which competition may occur, and so many different solutions to the problem of living in the soil are found by the large variety of organisms which have evolved.

The Major Groups of Soil Organisms

The micro-organisms found in soil are largely a reflection of the methods used to detect them. A detailed discussion of methods will be given in Chapter 3, but it is relevant here to consider a few general problems.

Firstly, the various groups of micro-organisms are so nutritionally diverse that no one isolation medium will suffice to isolate them. Secondly, comparisons of numbers of fungi and actinomycetes with numbers of bacteria developing on these media are meaningless since the units being counted are so dissimilar—spores and hyphae on the one hand, single cells or clumps of cells on the other. Thirdly, counts of micro-organisms do not necessarily indicate the relative importance of a group of organisms, since it is their rate of growth which determines their role in various processes. Thus, organisms may exist in a dormant or active state, and if active may multiply at vastly different rates.

Even within a closely related group of organisms, quite diverse ideas on the species present may be obtained with alternative methods of detection. Warcup[541] used two methods to isolate fungi, one of these (the dilution plate count) being selective for fungi present in soil as spores, and the other (the hyphal isolation technique) for fungi present as mycelium. The fungi most frequently isolated by these two methods are listed in Table 1. The lists of fungi obtained are markedly different since heavily sporing genera like *Penicillium* were

Table 1. *Fungi isolated from a wheatfield soil by two different methods*[541]

	Number of colonies developing	
	from spores (plate count)	from hyphae (direct isolation)
Penicillium spp.	5,499	0
Scopulariopsis sp.	476	0
Cladosporium herbarum	476	0
Rhizopus arrhizus	367	1
Mucor sp.	102	0
Fusarium oxysporum	223	16
Thielaviopsis sp.	81	1
Mortierella sp.	119	27
Sterile isolates	23	793

Table 2. *The general characteristics of the major groups of the soil microflora*

	Form and size	Cytology	Major components of cell wall	Methods of reproduction	Nutrition
Viruses of bacteria (phage)	Particles, often with head-like region (0·05–0·10 μm diam.) and tail-like region (up to 0·2 μm long)	DNA molecule with protein coat	—	Formation of new particles in host cell	Use host cell resources and synthetic pathways
Viruses of plants	Spherical to elongated particles (0·02–0·3 μm)	RNA molecule with protein coat	—	Formation of new particles in host cell	Use host cell resources and synthetic pathways
Bacteria	Single cells (0·5–2·0 μm $\times$ 1·0–8·0 μm) or loose associations of cells	Prokaryotic cells, lacking nuclear membrane, endoplasmic reticulum and mitochondria	Polysaccharides of sugars and amino sugars (e.g. galactose, glucosamine, muramic acid). Amino acids (e.g. diaminopimelic acid). Lipopolysaccharides Lipoproteins	Binary fission. Genetic exchange by transformation, transduction and conjugation	Heterotrophic or autotrophic, using chemical or light energy. O_2 not evolved in photosynthesis

Actinomycetes	Branched filaments (0·5-2·0 μm diam.)	Prokaryotic cells, lacking nuclear membrane, endoplasmic reticulum and mitochondria	Polysaccharides of sugars and amino sugars (e.g. galactose glucosamine muramic acid). Amino acids (e.g. diaminopimelic acid)	Conidia by fission of filaments. Fragmentation of filaments	Heterotrophic. A few autotrophic using chemical energy
Cyanophyceae (blue-green algae)	Single cells, loose associations of cells or filaments (2·0-5·0 μm diam.)	Prokaryotic cells, lacking nuclear membrane, endoplasmic reticulum and mitochondria	Polysaccharides of sugars and amino sugars (e.g. muramic acid). Amino acids (e.g. diaminopimelic acid)	Binary fission. Fragmentation of filaments	Autotrophic, using light energy. O_2 evolved in photosynthesis
Algae	Single cells, associations of cells or filaments (3·0-50 μm diam.)	Eukaryotic cells, with nuclear membrane, endoplasmic reticulum and mitochondria	Cellulose. Pectin and silicon in some	Sexual reproduction. Asexual reproduction by spores, fission or fragmentation	Autotrophic, using light energy. O_2 evolved in photosynthesis
Fungi	Filaments or single cells (3·0-50 μm diam.)	Eukaryotic cells, with nuclear membrane, endoplasmic reticulum and mitochondria	Chitin or cellulose. Glucan, mannan	Sexual reproduction Asexual reproduction by spores, fission, fragmentation or budding	Heterotrophic

Table 3. *The major features of the groups of fungi found in soil*

Group	Thallus form	Asexual reproduction	Sexual reproduction	Genera found in soil
Myxomycetes (true slime moulds)	Acellular, multinucleate plasmodium	Non-motile spores	Fusion of flagellate gametes	*Physarum*
Chytridiomycetes	Unicellular, globose to filamentous	Uniflagellate zoospores	Fusion of uni-flagellate gametes	*Allomyces* *Rhizophydium*
Oomycetes	Unicellular to aseptate filaments	Biflagellate zoospores	Fusion of gametes to give oospores	*Pythium* *Saprolegnia*
Zygomycetes	Aseptate or septate filaments	Non-motile sporangiospores	Fusion of gametangia to give zygospore	*Absidia* *Mortierella* *Mucor* *Rhizopus* *Zygorhynchus*
Ascomycetes	Septate filaments or unicellular (yeasts)	Non-motile conidia	Ascospores produced by meiosis of diploid nucleus. Held inside ascus	*Chaetomium* *Gymnoascus* *Sordaria* *Saccharomyces*
Basidiomycetes	Septate filaments	Non-motile conidia	Basidiospores produced by meiosis of diploid nucleus. Held on basidium	*Boletus* *Corticium* *Marasmius* *Omphalina* *Tricholoma*
Deuteromycetes (Fungi Imperfecti)	Septate filaments	Non-motile conidia	Absent	*Arthrobotrys* *Aspergillus* *Cladosporium* *Fusarium* *Gliocladium* *Penicillium* *Trichoderma*
Mycelia Sterilia	Septate filaments	Absent	Absent	*Rhizoctonia*

Table 4. *The major features of the groups of algae found in soil*

Group	Thallus form and major cell wall components	Asexual reproduction	Sexual reproduction	Pigments	Stored photosynthate	Genera found in soil
Chlorophyceae (green algae)	Unicellular, colonial or filamentous. Cellulose	Zoospores, division of unicells, or fragmentation of filaments	Fusion of motile or non-motile gametes	Chlorophyll a, b. α-, β-, y-carotenes. Xanthophylls	Starch	*Chlamydomonas* *Chlorella* *Pleurococcus* *Ulothrix* *Zygnema*
Bacillariophyceae (diatoms)	Unicellular, colonial or filamentous. Pectin Silicon dioxide	Cell division	Fusion of naked gametes (motile and non-motile)	Chlorophyll a, c. β-carotene. Xanthophylls	Oil Chrysolaminarin	*Achnanthes* *Navicula* *Pinnularia*
Xanthophyceae (yellow-green algae)	Unicellular, or filamentous. Pectin. Few with silicon dioxide	Zoospores, division of unicells, or fragmentation of filaments	Fusion of motile or non-motile gametes	Chlorophyll a, e. β-carotene. Xanthophylls	Oil Chrysolaminarin	*Botrydiopsis* *Heterococcus* *Vaucheria*
Cyanophyceae (blue-green algae) unlike true algae, have prokaryotic cells	Unicellular, colonial or filamentous. Polysaccharides. Peptides	Binary fission, fragmentation of filaments, production of thick-walled spores	Absent	Chlorophyll a. Phycocyanin. Phycoerythrin β-carotene. Xanthophylls	Cyanophycean starch	*Anabaena* *Chroococcus* *Nostoc*

Table 5. *The major features of some genera of bacteria found in soil*

Genus	Cell shape	Spores	Motility	Gram stain reaction	Oxygen requirement	Other features
EUBACTERIALES						
Arthrobacter	Rods, later forming cocci	–	–	Variable	Aerobic	Form cystites. Little reaction in many of the usual biochemical tests
Azotobacter	Rods or yeast-like cells	cysts	+ or – peritrichous flagella	–ve	Aerobic	Fix atmospheric nitrogen and grow best on nitrogen deficient media
Bacillus	Rods	+	+ or – peritrichous flagella	+ve	Aerobic	Fermentative metabolism. Usually proteolytic. Rarely pigmented
Clostridium	Rods	+	+ peritrichous flagella	+ve	Anaerobic	Fermentative metabolism. Often proteolytic. Some fix nitrogen
Micrococcus	Cocci	–	–	+ve	Aerobic or microaerophilic	Fermentative metabolism. Often tolerate high osmotic pressure. Pigmented
Nitrosomonas	Rods	–	+ or – polar flagella	–ve	Aerobic	Oxidise ammonia to nitrite
Pseudomonas	Rods	–	+ or – polar flagella	–ve	Aerobic	Oxidative metabolism. Often produce fluorescent pigments
Rhizobium	Rods	–	+ or – peritrichous or polar flagella	–ve	Aerobic	Glucose utilised without much acid formation. Form nodules with legumes and fix nitrogen
MYXOBACTERIALES						
Chondromyces	Flexible rods, blunt or pointed	Cysts on end of coloured stalk	Gliding motility	–ve	Aerobic	
Cytophaga	Flexible rods, often pointed	–	Gliding motility	–ve	Aerobic	Difficult to distinguish from pigmented eubacteria

predominant on soil dilution plates, while fungi developing from pieces of hyphae picked out of soil remained sterile in culture. In subsequent work, Warcup and Talbot[545-547] induced many of these to fruit and found that they were mostly Basidiomycetes and Ascomycetes. One proved to belong to a new genus, eight were new species and nearly all of them were fungi which had not previously been detected in soil.

Faced with these considerations, it will be readily appreciated why we have not attempted to prepare a comprehensive list of 'typical' soil-inhabiting micro-organisms. Rather than do this, we have tried to show how micro-organisms are adapted to life in the soil in respect of growth and spread, resistance to inimical environmental factors, variation, growth rate and nutrition. However, for those who are unfamiliar with the general properties of micro-organisms, we have summarised their main diagnostic features in a series of tables. Table 2 shows the main characteristics of the fungi, algae, bacteria, actinomycetes and viruses, while Tables 3, 4, 5 and 6 deal with some of the more frequently isolated forms within these groups.

Table 6. *The major features of some genera of actinomycetes found in soil*

Genus	Main features
Micromonospora	Filaments do not grow above medium. Single spores produced in and on surface of medium. Colonies rather slow growing on most media
Nocardia	Filaments unstable, fragmenting into bacteria-like units. Filaments usually not growing above medium and spores rarely produced
Streptomyces	Long chains of spores formed on filaments growing above the medium. Species very numerous in soil and many produce antibiotics
Streptosporangium	Spores formed in sporangia or in chains on the filaments above the medium. Colony appearance similar to *Streptomyces*
Thermoactinomyces	Single spores formed on filaments above and within the medium. Spores heat-resistant. All species are thermophilic

Spread of Micro-organisms in Soil

(a) *Growth Patterns*

Substrates for growth are present in many forms in the soil and micro-organisms are structurally adapted to use them in many different ways. At one extreme, we find small localised colonies on particulate substrates, and at the other, diffuse widespread colonies not particularly associated with any definable substrate. Burges[59] distinguished several groups of soil fungi on the basis of their growth patterns, and in the ensuing discussion his ideas have been extended to include other soil organisms: some of his groupings have been retained, although different names have been used to describe them.

(*i*) *Non-migratory, Unicellular Pattern* (Plate I (a)) Unicellular yeasts and non-motile bacteria, e.g. *Arthrobacter*, usually exist in soil as single cells or in the form of compact colonies, usually containing 10-20 cells, fixed on to soil particles. In regions where nutrients are plentiful, larger colonies may form. Such organisms do not appear to be specially adapted for spreading through the soil, except that their small size may allow them to be moved passively through the soil pores (see p. 12). Many bacteria that are motile in culture probably exhibit a non-migratory growth pattern in soil, where conditions favouring flagellum production may not occur. It is difficult, therefore, to distinguish this growth pattern from the next one to be described.

(*ii*) *Migratory, Unicellular Pattern* (Plate I (b)) Some bacteria, certain chytrid fungi, diatoms and some blue-green and green algae possess limited powers of movement, at any rate in laboratory culture, e.g. *Pseudomonas viscosa* which can swim about 5-6 cm in 8 hours in water. Conclusive proof of active movement in the soil is lacking, mainly because it is impossible to distinguish it from passive movement through soil However, Zvyagintsev[576] has observed the interchange between solid surfaces and water films in simulated soils. Brown, Jackson and Burlingham[55] showed that *Azotobacter* could travel along root surfaces but hardly at all through soil. Also Starkey[475] showed that soyabeans inoculated with rhizobia did not transmit these bacteria to adjacent, uninoculated plants and concluded that rhizobia remained restricted to the root region for at least several weeks. Thus, if movement does occur in the soil, it is probably associated with continuous surfaces, e.g. along roots, rather than from one surface to another, e.g. from one soil particle to another.

(*iii*) *Plasmodial Pattern* Fungi belonging to the Myxomycetes do not possess hyphae, but consist of naked masses of protoplasm which move in an amoeboid fashion until eventually they are transformed into a mass of walled spores which form a complex fruiting body. These fungi are able to migrate over the soil surface from one substrate to another, but probably do not do so below the surface. The myxobacteria may have a similar growth pattern, but their ability to spread will be somewhat less.

(*iv*) *Restricted Hyphal Pattern* (Plate I (c)) Fungi such as *Penicillium* and actinomycetes, e.g. *Streptomyces*, form dense hyphal growths on and within small pieces of substrate. They form large numbers of spores which are generally non-motile and which may fall off into the soil, but there is little or no extension of the mycelium into the soil pores. Many of the so-called sugar fungi (see Chapter 4) probably have growth patterns of this type and between exhausting one substrate and being presented with another will remain dormant as spores[543]. Obligate parasites, e.g. certain root-infecting fungi, also show restricted growth in soil since they are by definition confined to living host tissues. A few of them can form motile spores, but their range is so small that probably their chief function is to locate infection sites on the root.

Filamentous algae may also have restricted growth, but in this case, since light and not organic matter is the source of energy, they grow in concentrated form at the soil surface.

Burges[59] suggested that this type of growth pattern might be correlated with the ability of fungi to produce antibiotics, allowing them to dominate restricted environments. The inclusion of actinomycetes in this group emphasises the point.

(*v*) *Locally Spreading Hyphal Pattern* The fungus *Mucor ramannianus* may colonise a substrate and form non-motile aplanospores, or it may spread into the surrounding soil where it forms chlamydospores (p. 17). The spreading mycelium can then disintegrate, leaving the chlamydospores unassociated with the original substrate. Whether this habit is typical of chlamydospore-forming fungi is not known, but it may also be found in some mycorrhizal fungi, e.g. *Endogone* (see p. 128).

(*vi*) *Mycelial Strand or Rhizomorph Pattern* (Plate I (d)) The hyphae of many basidiomycetes may become entangled to form a mycelial strand which can pass through the soil for long distances from one substrate to another. Sometimes the hyphae become organised into root-like structures called rhizomorphs which often have a central core of living hyphae, surrounded by a protective sheath of thickened hyphae. Rhizomorphs grow faster than single hyphae and have considerable penetrative powers which enable them to colonise hard, woody substrates. Many parasitic and symbiotic fungi possess rhizomorphs or mycelial strands, allowing them to travel from one host to another.

Basidiomycetes may also form non-motile spores, especially in fruit bodies produced above the soil, and these may be spread by air currents and germinate to produce mycelium at some distance from the original area of growth.

(*vii*) *Diffuse Spreading Hyphal Pattern* (Plate II (a)) Examination of soil often shows isolated fungal hyphae which grow randomly through the soil and which are not associated with particulate substrates. Burges[59] has suggested that *Zygorhynchus* belongs to this group and that it grows on the nutrients in the soil solution.

Diffuse, spreading growth is also exhibited by the fairy-ring fungi, although these forms grow in a definite direction, i.e. outwards from a central area. As they migrate outwards, the central parts may die back, but not before many chemical and microbiological changes have taken place in the soil[96].

(*b*) *Spread of Microbial Fragments*

Many colonies of micro-organisms can fragment into small mycelial pieces, spores or unicells which remain viable and which may be moved through the soil by outside influences, rather than by active growth or movement.

Propagules may be spread outside the soil through the air, but apart from those fungi which have aerial or soil fruiting bodies with special spore discharge mechanisms, soil organisms only become dispersed in the air incidentally when soil is blown by wind or disturbed in other ways. Thus Hirst and Stedman[210] suggested that litter organisms might be dispersed when raindrops hit the soil surface.

Inside the soil, propagules may be moved by percolation of water or by animals. Percolation of water is responsible chiefly for downward movement of fragments and spores, but for efficient movement to take place the fragments should be small and regularly shaped. Large propagules will not be able to pass through the soil pores and irregularly shaped ones may become jammed across the pores. Spores, such as conidia, and bacterial cells are best fitted for this type of movement, while fragments of mycelium are poorly adapted. Propagules dispersed by water must also have suitable surface properties and Burges[57] has shown that easily wetted spores are more readily moved through soil by water than spores with hydrophobic surfaces. The electrical charge on the surface of the propagules is also important. Most soil particles are negatively charged; so are most micro-organisms, and as a result of this the micro-organisms are not absorbed on to the particle surfaces and can pass through the soil pores. However, some soils contain positively charged particles, e.g. the B horizons of soil profiles which contain trivalent cations like iron, and here mobility of propagules is poor because of adsorption.

Movement of propagules on or in the bodies of small animals is also important. Some fungal and actinomycete spores are highly ornamented with hairs, warts and spines. It is possible that these ornamentations help the spores to be picked up on the bodies of small soil animals, many of which also have complex, spiny surfaces. It has also been suggested that ornamented spore surfaces may prevent downward movement due to percolation of water. Unlimited downward movement might not be advantageous to an organism for it could be carried away from the substrates necessary for its growth.

Micro-organisms are also eaten by soil animals, sometimes deliberately, e.g. by mites browsing on fungi decomposing leaf litter, but more often incidentally during the ingestion of organic matter. These micro-organisms may be passed through the animal body and thus arrive in new environments while still viable. For this to happen, it is essential that the cell wall is not attacked by enzymes in the animal gut, and it is noteworthy that some fungal cell walls cannot be decomposed unless attacked by a variety of different enzymes.

Ingestion of organic matter and egestion of faeces also results in transport of nutrients, bringing them to micro-organisms which have no means of transport of their own. Soluble nutrients in the soil solution can be carried to static micro-organisms, in which case microbes remaining in one site will benefit by waiting for food to arrive and by not expending energy during migration.

Resistance of Micro-organisms to Extreme Environmental Conditions

Since many soil micro-organisms have little ability to spread, it is important that they develop ways of resisting unfavourable environmental conditions such as desiccation, high salt concentrations, high temperatures, presence of toxins, anaerobiosis and lack of substrate. Resistance may be achieved by modifications to the vegetative structure of the organism or by the production of specialised structures like spores and cysts.

(*a*) *Vegetative Structures*

It is thought that many vegetative cells and pieces of mycelium can become dormant and survive for quite long periods of time. Bacterial cells may form a protective capsule of polysaccharides and polypeptides which can imbibe water and nutrients from the environment and discourage predation by protozoa. Spherical, non-motile resting cells may also be formed, but why they should be resistant is not known. Assessing the contribution of such cells to the survival of the organisms is difficult since it is hard to distinguish survival of dormant, vegetative propagules from low but continuous growth of a small number of cells, especially when this takes place on a localised basis in micro-environments.

The vegetative mycelium of some fungi may be resistant to attack during periods of inactivity by virtue of their thick, often pigmented walls (see p. 104). Mycelium may also be modified to form special bodies called sclerotia. These are compact, globular masses of closely packed, repeatedly branching hyphae, the outer surfaces of which consist of very thick-walled cells impregnated with melanin-like pigments (Plate II b)). They may survive in soil for several years and for that reason are often formed by plant pathogenic fungi so that they may survive in the absence of host plants. Other massive fungal structures such as rhizomorphs may also aid survival, as well as assisting in the spread of an organism.

(b) Spores

Resistant spores of many different types are formed by a wide variety of soil organisms. In general, they differ from spores involved in dispersal as they are more massive and thick-walled, although this distinction is by no means clear-cut. In this discussion we shall consider several common spore types including bacterial endospores, asexual fungal spores, sexual fungal spores and akinetes of blue-green algae.

(i) Bacterial Endospores Cells of the bacteria *Bacillus* and *Clostridium* and a few others can produce endospores within their cells (Plate III (a)). These spores are extremely resistant to high temperatures, desiccation and radiation and are more resistant to inhibitory compounds, e.g. antibiotics and alcohols, than are the vegetative cells.

Laboratory experiments have concentrated on their heat resistance and it has frequently been observed that these spores can resist temperatures of 80-100°C for periods of half an hour or more. It is not clear how this could affect their ability to survive in nature, for exposure to such high temperatures for any length of time will only occur in thermal regions and after forest or grass fires. Probably resistance to desiccation injury and toxic substances are more important and in this connection, Elwan and Mahmoud[125] have shown that spore-forming bacteria predominate in desert soils, while Sneath[457] found that almost all the organisms isolated from soil on the roots of plants kept in herbaria for up to 300 years were species of *Bacillus*. Bacilli are also common in acid forest soils where toxic humic substances and high concentrations of iron and aluminium ions are present.

The reasons for the resistance of endospores are not fully understood, but one widely favoured view is that the internal core of the spore is maintained in an anhydrous state, possibly by the compressive contraction of the outer parts of the spore[268]. Proteins in such an anhydrous state might be resistant to denaturation in a way analogous to those in freeze-dried cells.

Park[349] suggested that the conditions for endospore formation might not be encountered in soil, but Sussman and Halvorson[498] have pointed out that endospores are usually formed as nutrients are being depleted, following a period of vegetative growth in nutrient-rich conditions. This fluctuation in nutrient conditions is precisely what one would expect to find in soil. Furthermore, pasteurisation experiments and direct observation of soil[209] have clearly shown the presence of spores.

(*ii*) *Spores of Actinomycetes and Blue-green Algae* Prokaryotic organisms other than bacteria also produce a variety of spores, but their role in resistance to unfavourable environmental conditions is less clearly established.

Actinomycetes such as *Streptomyces* produce thin-walled conidia which have hydrophobic properties not unlike those of some fungal conidia. They are generally supposed to be resistant to desiccation[318] and can survive for long periods in dried soil. Szabó, Marton and Varga[502] also showed that they were more resistant to heat than vegetative mycelium and could survive partial heat sterilisation of soil (see also Chapter 3). A recent investigation of *Thermoactinomyces vulgaris* has shown that its 'conidia' are very similar in fine structure and properties to the endospores of *Bacillus* for they are markedly heat resistant and contain a high percentage of dipicolinic acid[98].

Blue-green algae also produce spores of many different types. Unfortunately, most work has been carried out on aquatic forms and little is known of the mechanism of survival of terrestrial forms[136]. Members of the Nostocaceae and Rivulariaceae produce resting spores called akinetes. These are spherical, oblong or cylindrical cells which contain food reserves of cyanophycin, but which lack photosynthetic pigments. They have a firm wall composed of two layers, the outer one being thick, yellow brown in colour and frequently ornamented. Often, they are produced in chains from a row of vegetative cells, especially during periods of nitrogen deficiency. They are resistant to desiccation and extremes of temperature but require light or abundant supplies of carbohydrate for their formation. Some forms, e.g. the Scytonemataceae and Stigcnemataceae, produce hormocysts which are not unlike many-celled akinetes and which are thought to be adaptations to terrestrial conditions. *Tolypothrix penicillata*, another tcrrestrial form, can survive as isolated threads with a thick yellow sheath. On the resumption of activity, thesc quickly form trichomes by rapid cell division.

Finally, blue-green algae, which can fix atmospheric nitrogen, often form heterocysts[482], e.g. members of the Nostocales and Stigonematales. These are larger than vegetative cells and have a thick two-layered wall of cellulose and pectic materials[136]. The cytoplasmic membrane of these heterocysts may remain continuous with that of adjacent cells because of a small pore present at each end of the cyst. The protoplast eventually disappears and germination and revitalisation have rarely been observed, although Wolk[572] has recorded germination in the presence of ammonium nitrogen. It has been suggested that materials may be transferred to the heterocysts during periods of nitrogen starvation[134], thus helping the filament to survive. However, there is now a considerable body of evidence suggesting that the heterocysts are the sites of nitrogen fixation[132] (see also p. 136).

(*iii*) *Fungal Spores* Fungi produce a variety of spores (Plate III (b), (c) and (d)) including chlamydospores, conidia, sporangiospores, oospores, zygospores, ascospores and basidiospores. One of the important functions of these spores is to lie dormant until substrates are available. During this waiting period, environmental conditions may be unfavourable and so these spores must be able to resist such things as desiccation and toxic substances.

There is surprisingly little evidence concerning the relative importance of different spore types in survival, for generally survival time has been assessed by recovering fungi from samples without knowledge of their form. Also, little is known about the conditions needed to form spores in soil although we do know that various fungi require different environments for sporing. Griffin[172] has shown that the conidia of *Curvularia* form only in air-filled pores while *Fusarium culmorum* form mainly in water films.

Sussman[497] has reviewed the information available on the longevity of spore types and Table 7 shows a selection of the evidence he brought

Table 7. *Longevity of fungal spores and sclerotia in soil*

Organism	Spore type	Longevity
Phytophthora infestans	Sporangia	9-10 weeks[497]
Helminthosporium sativum	Conidia	>20 months[497]
Trichoderma viride	Conidia	> 1 year[67]
Cochliobolus sativus	Conidia	*c.* 50 days[343]
Fusarium oxysporum	Conidia	< 8 weeks[440]
Fusarium oxysporum	Chlamydospores	> 8 weeks[440]
Plasmodiophora brassicae	Resting spores	>10 years[497]
Peronospora destructor	Oospores	> 8 years[497]
Aphanomyces euteiches	Oospores	10 years[497]
Fomes annosus	Basidiospores	<18 months[497]
Claviceps purpurea	Sclerotia	2 years[497]
Verticillium dahliae	Sclerotia	14 years[497]
Sclerotium rolfsii	Sclerotia	<1->60 months[497]

together, as well as the results of some other workers. Spores with thick or heavily pigmented walls seem to be particularly resistant and several investigators have shown that relatively thin-walled spores are drastically affected by desiccation, e.g. the conidia of *Penicillium* and sporangiospores of *Rhizopus*[542], while Park[346, 347] has shown that many conidia are readily decomposed by the soil microflora unless they germinate and almost immediately form chlamydospores.

Chlamydospores are probably among the most important of asexual spores in relation to survival. They are thick-walled, have dense oily contents, limited powers of germination and considerable longevity. They may develop from conidia or from well-nourished mycelium, often when the environment is depleted of essential nutrients, e.g. hexose sugars[194]. Carlile[68] also noted that a low carbon : nitrogen ratio favoured chlamydospore formation in *Fusarium oxysporum* f. *gladioli*. Sequeira[440] made similar observations in a study of *Fusarium oxysporum* f. *cubense* for he found that no chlamydospores formed when the carbohydrate content of soils was high. When no chlamydospores were formed, the fungus began to disappear from the soil, mainly because both conidia and hyphae were lysed by a mixture of unidentified soil bacteria in as little as 12-15 days. Table 8 shows the change in

Table 8. *Changes in the spore populations of* Fusarium *present on slides buried in soil*

Spore type	Percentage before burial	Percentage after twelve days burial
Microconidia	94·1	28·5
Macroconidia	5·9	2·8
Chlamydospores	0·0	68·7

proportion of different spore types on slides buried in soil, a change which resulted in conidia being entirely eliminated from the soil after 8 weeks.

Not all chlamydospores are so resistant to lysis. The albino mutant chlamydospores of *Thielaviopsis basicola* can be lysed quickly, although the normal melanin-pigmented ones are not, suggesting that complexes between the wall components and melanin are involved in the resistance mechanism.

Many asexual spores have comparatively low water contents, when compared with the vegetative mycelium which formed them. Mazur[314] quotes water contents of 0·06-0·3 g per g dry weight for a variety of spores. It is only when spores are present that many fungi will survive freeze drying[207]. This suggests that spores are able to resist desiccation, but why this is so is not known. Dehydration involves three simultaneous processes, removal of water, concentration of ions and low molecular weight substances, and a decrease in distance separating macromolecules. Sussman[497] suggests that some of the characteristic components of fungal spores such as trehalose and mannitol, and the dipicolinic acid in bacterial spores, may replace water as stabilising agents for macromolecules like DNA.

Low water contents may be maintained in thick-walled spores because the rigid wall cannot expand in response to a volume increase that would follow water uptake. Mazur[314] estimates that the walls of a spore containing 20 per cent water might have to withstand pressures of 876 atmospheres in an environment of 95 per cent relative humidity. Among the most thick-walled spores, with abundant food reserves, are certain of the sexual spores, e.g. oospores and zygospores.

Other sexual spores, whilst comparatively thin walled, are enclosed in massive protective structures, e.g. ascospores inside asci formed within perithecia or cleistothecia which isolate the spores from the environment. Very little is known about the structure and chemical composition of sexual spores so that it is difficult to say why they are resistant.

Variation in Soil Micro-organisms

Since soil micro-organisms grow in a varied and fluctuating environment, they must be capable of adaptation to new conditions. Those organisms which can grow and reproduce best under these conditions will make greater genetic contributions to each succeeding generation and, in the process, certain genes and gene combinations will be favoured[361]. The original source of variation is mutation, but mutation of any particular gene occurs only rarely, perhaps only once during the production of a million or a hundred million new nuclei. However, micro-organisms grow very rapidly, at least when presented with fresh substrates, and so mutant forms may arise quite frequently. In higher organisms, mutant genes are not always evident for they may be masked by non-mutant alleles in diploid nuclei. However, many micro-organisms are haploid for much of their life and so the effect of the mutant gene is immediately expressed. If the mutation favours growth and reproduction under the prevailing environmental conditions, then it will survive and oust other less successful forms; otherwise, it will die out in the face of rigorous competition. Soil organisms which have extended haploid growth phases include the bacteria, actinomycetes and blue-green algae, and most phycomycete and imperfect fungi. Soil organisms with extended diploid phases include some of the yeasts, some myxomycetes, and a few Phycomycetes, e.g. certain *Pythium* spp. Basidiomycetes have hyphae consisting of cells containing two genetically distinct nuclei which only fuse prior to spore formation (see below).

The haploid nature of many soil organisms is disadvantageous in some respects since the ability to store variability from one generation to another is limited. This is perhaps a more serious problem for bacteria rather than fungi, for the latter group have evolved other ways of storing variability. Thus a fungal hypha may contain several genetically distinct haploid nuclei in a common cytoplasm, a phenomenon termed heterokaryosis. By changing the proportions of these different nuclei, the growth and reproduction of the organism can be altered to suit the environment. However, the existence of any particular heterokaryon is short, at any rate in fungi which produce uninucleate spores where heterokaryons must be re-established in each generation by mutation, or by anastomosis between hyphae growing from different spores[400]. Jinks[236] has suggested that heterokaryon formation is favoured in certain penicillia by the dispersal of intact chains of spores. In view of the restricted growth patterns of some penicillia in the soil, this may be particularly important. Jinks[237] has also suggested that genetic variability might be stored in an analogous way by genetically distinct cytoplasmic determinants forming heteroplasmons.

New variants may also be produced by reassortment of the available hereditary material. This is achieved in many higher organisms by sexual recombination. Some micro-organisms also exhibit sexual processes, but a number of other recombination methods are known. For instance, in bacteria, molecules of DNA liberated from cells may be acquired by other cells, which then develop new heritable properties. This process, known as transformation, is only known from experiments with laboratory cultures and it seems unlikely that the precise conditions required for it to occur would be found very often in soil. Nevertheless, it has been suggested that a massive release of DNA from bacteria in decaying root nodules might influence bacteria growing in the rhizosphere[131]. Other processes involving the transfer of genetic material from one bacterium to another include transduction, which involves the transfer of part of the bacterial chromosome from one bacterium to another while it is associated with a bacterial virus, and conjugation, where a substantial part of a bacterial chromosome is passed from one cell to another via a conjugation tube. Like transformation, these processes have only been recorded in laboratory cultures and it seems that they are also unlikely to occur in soil, except in regions of high nutrient status where multiplication is rapid, e.g. on root surfaces. From a genetic viewpoint, these recombination processes differ from true sexual methods because the genetic contributions of the partners are almost always unequal, and in consequence, the offspring usually resemble one parent more closely than the other.

Fungi often possess normal methods of sexual reproduction, involving the fusion of nuclei in gametes, gametes and gametangia, gametangia, and vegetative hyphae to form oospores, zygospores, etc. Fusion is followed by meiosis, sometimes almost immediately and sometimes after the growth of a diploid phase, leading to the production of zoospores, ascospores and basidiospores. Again, the evidence that sexual spores are formed in soil is limited, although plant pathogens like *Pythium* form oospores in decaying plant tissues. Many of the thick-walled, sexual spores are difficult to germinate in the laboratory, suggesting that they are important as survival structures. Other thin-walled sexual spores are light and act as dispersal agents in the atmosphere above the soil, e.g. ascospores and basidiospores. When basidiospores germinate, they give rise to a haploid mycelium. This grows and eventually fusion takes place between different hyphae. The component nuclei become associated in pairs within each cell of the new mycelium and divide along with the cell. Such mycelia are termed dikaryons and their importance lies in the fact that fusion of hyphae and fusion of nuclei are separated; since the component nuclei divide repeatedly between these events, the number of spores and possible genetic combinations produced per sexual fusion is increased enormously[388]. Only dikaryotic mycelia can give rise to fruiting bodies, proving that hyphal fusions must take place within the soil system.

Sometimes sexual spores are formed only when fusion occurs between sexually different strains (heterothallism), but more frequently fusion takes place between sexually identical strains, often between nuclei from the same hyphal system (homothallism). It might be thought that homothallism is of little value genetically since it usually involves the fusion of genetically identical nuclei; nevertheless, it is the most common sexual condition found in fungi, probably because fusion between different strains is unlikely in environments like soil where hyphae may be spatially separated by comparatively great distances. As Raper[388] points out, fungi seem to be adapted to produce quantity rather than quality in their sexual spores. Possibly, mutations occur within the mycelium sufficiently often to permit reassortment of genetic information in homothallic systems.

We have already noted that some fungi do not possess any sexual mechanisms at all. This does not mean that recombination of genetic material does not occur, for it has been shown that in some Fungi Imperfecti, nuclei in heterokaryons may occasionally fuse and recombination occur during mitosis rather than meiosis. Pontecorvo[372] has referred to this process as the parasexual cycle and has shown that in *Aspergillus nidulans*, which has a sexual cycle as well, recombination by the parasexual method is about five hundred times less frequent than in the sexual cycle; in forms lacking sexual reproduction, e.g. *Penicillium chrysogenum*, parasexual recombination occurs more frequently. So far, it has not been shown whether mitotic recombination occurs in nature or even in simulated natural environments, but there seems no reason to doubt that it does.

Finally, variation in micro-organisms can be caused by changes in cytoplasmic genetic factors. These may be wholly cytoplasmic, or they may alternate between cytoplasmic and chromosomal conditions. These latter factors are termed episomes and have been described in a number of bacteria. They have a remarkable similarity to temperate bacterial viruses in the sense that they can spread epidemically through populations. It is known that bacterial viruses occur in soil and are presumably transmitted from one cell to another, suggesting that spread of episomes in populations could be a further important method for perpetuating advantageous variants.

Microbial Growth and Enzyme Production

Micro-organisms are remarkable for the diversity of energy sources they may employ. Light, oxidation of inorganic chemicals and dissimilation of almost every conceivable organic substance can be used by one micro-organism or another under appropriate environmental conditions. This in itself is a remarkable adaptation to life in the soil and will be considered in detail in subsequent chapters. A further adaptation to growth in the soil is the frequent occurrence of individual organisms which can change the balance of enzymes in the cell, enabling them to live in different ways, e.g. aerobically and anaerobically, autotrophically and heterotrophically, or on a variety of different carbon and nitrogen sources. The ability to produce the necessary enzymes is governed genetically, but although an organism has an enormous genetic potential, it would be wasteful if all of this potential were to be expressed simultaneously. This would result in the production of small amounts of many enzymes (not all of which would be able to function since their substrates might be absent) and relatively slow growth rates. Therefore, it is not surprising that mechanisms controlling the synthesis and activity of enzymes have been evolved, the guiding principle of which is the attainment of maximum growth rates under given nutrient conditions.

(a) *Enzyme Induction, Repression and Inhibition*

Control of enzyme synthesis has been investigated principally in organisms not normally thought of as soil inhabitants, e.g. *Escherichia coli*. In *E. coli*, clear evidence has been obtained that the presence of a substrate, e.g. lactose, can induce the formation of an enzyme, e.g. β-galactosidase, that otherwise would not have been formed. The work of Jacob and Monod[226] and others has shown how this induction might take place. The following account is taken from a recent review by Clarke and Lilly[88] who have summarised much of the information on the regulation of enzyme synthesis during microbial growth (see also Fig. 1).

In the absence of the appropriate substrate, a regulator gene (i) produces a repressor molecule which prevents the expression of structural genes (z, y and a) which, in turn, determine the enzymes β-galactosidase, β-galactosidase permease and thiotransacetylase. This is brought about by the repressor combining with an operator gene (o) which would otherwise allow transcription of the structural genes, production of messenger RNA and enzyme synthesis. However, when the substrate for the enzyme z is present, it combines with the repressor, allowing the operator to 'switch on' the sequence of structural genes determining enzyme synthesis. A further genetic element, the promoter (p), seems to be necessary for the initiation of transcription. Many variations of this scheme have been suggested for control of other enzymes in *E. coli* and for enzymes in other bacteria, but so far, similar control mechanisms have not been detected in fungi, although enzyme induction is well known[64].

When enzymes have been produced, there exists the possibility that they may become too active, in which case other enzymes may be robbed of their substrates. To guard against this, mechanisms controlling enzyme activity have been evolved. One such mechanism is feedback-inhibition in which the end-product of a series of reactions inhibits the activity of an enzyme in an early stage of its own synthesis. Feedback-inhibition has often been found to regulate amino acid biosynthesis, and may be caused by the combination of enzyme and end-product, resulting in an altered conformation and inactivation of the enzyme.

The end-product of a series of reactions may also prevent synthesis of one or more enzymes which catalyse earlier stages in its formation, a phenomenon termed enzyme repression. Repression of some enzymes may also be inhibited by carbon compounds, other than end-products of the type described. Thus glucose seems to inhibit many enzymes. It has been suggested that it is not glucose itself which causes the repression, but a compound derived from it which is also an intermediate of the metabolism of the substrate inducer[297]. Both enzyme

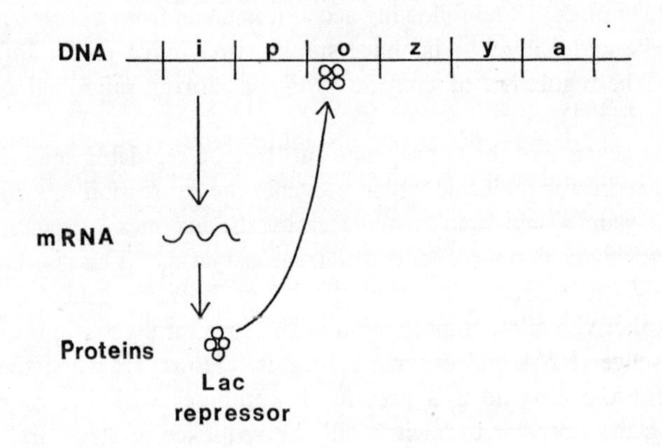

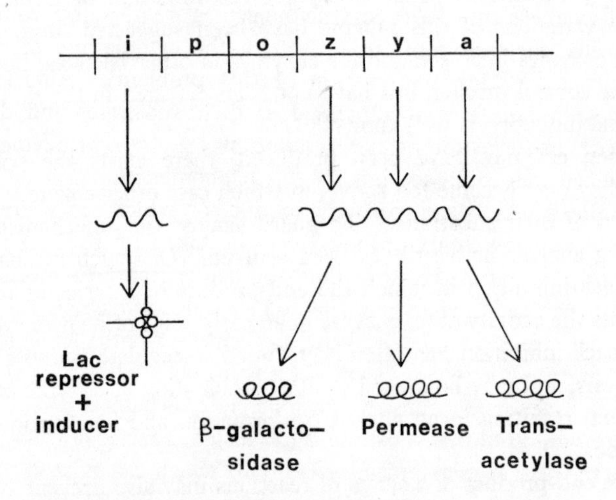

Fig. 1. *Model for the regulation of the lactose operon.* (a) *The regulator gene* i *produces a repressor molecule which combines with the operator* o *to prevent transcription.* (b) *An inducer combines with the repressor removing it from the operator, allowing transcription of genes* z, y *and* a. *The promotor* p *is essential for the initiation of transcription (Clarke and Lilly*[88])

(*Reproduced by permission of P. H. Clarke, the Society for General Microbiology and Cambridge University Press*) repression and feedback-inhibition have been recorded for enzyme systems in both bacteria and fungi.

(*b*) *The Importance of Enzyme Control in the Soil*

Mechanisms of induction and repression may be expected to be of considerable importance in environments like soil, where nutrients fluctuate in both quantity and quality. However, in the soil, it is common to find nutrients in an insoluble state or in low dissolved concentrations and so it is worth examining how some of the processes we have described are affected by these considerations.

If a substance is insoluble or if the cell is impermeable to it, it is difficult to see how it can act as an enzyme inducer. Pramer[376] has pointed out that both chitinase and keratinase are inducible enzymes and yet chitin and keratin are both insoluble. How, and where, they initiate enzyme synthesis is unknown, although it is possible that induction is brought about by small, more soluble molecules derived from these insoluble polymers[50]. When the enzymes are produced, they act extracellularly on the insoluble polymers, but it is still not clear how large enzyme molecules are liberated from intact living cells, for they must pass through the cytoplasmic membrane (see Lampen[260] for a discussion of this problem). Alternatively, cells may become closely adpressed to their substrates and dissolve them without formation of extracellular enzymes[466]. Enzymes built into the cell wall could account for this phenomenon.

Substances which are sparingly soluble, or which are soluble and yet present in low concentrations, may be taken up by bacteria and fungi and concentrated within them by active transport mechanisms termed 'permeases'. Each enzyme, e.g. β-galactosidase, is associated with a permease, e.g. β-galactosidase permease, both being induced by the same compounds. Permeases ensure that substrate is supplied to an enzyme at a relatively constant rate, whatever the external concentration, until that substrate is exhausted, a factor of clear importance in the near growth limiting conditions which can exist in the soil. Other transport systems are known which are constitutive, i.e. they are present whether their 'substrate' is or not.

(*c*) *Growth Rates and Starvation Conditions*

Much of the information we have discussed indicates that the microbial cell is a self-regulating system in which ability to attain maximum growth rate is all-important. It is not surprising, therefore, that soil microbiologists have rated ability to grow rapidly as one of the most important attributes of a highly competitive soil organism. Growth rates are usually assessed in laboratory culture where nutrients are plentiful, and it has often been assumed that the rates observed would bear the same relation to one another that they do in soil, where nutrients are scarce. This may not be so, and it is highly probable that various organisms respond in different ways to growth limitations.

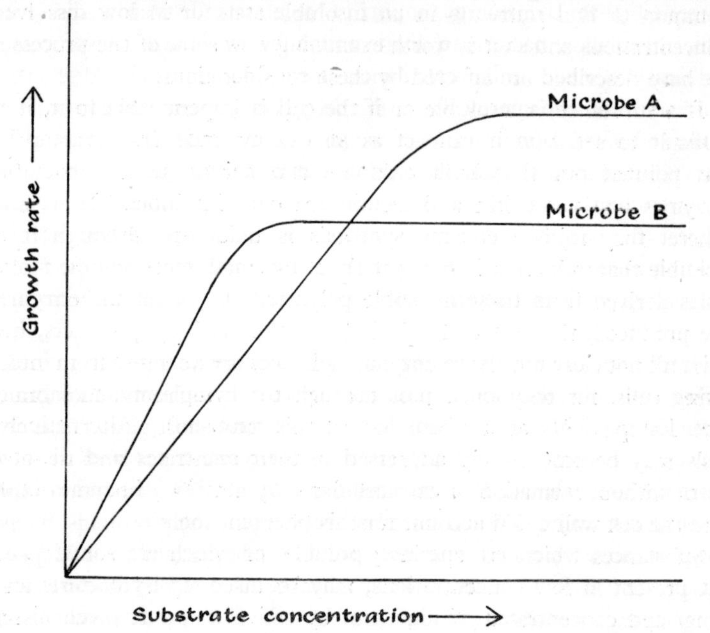

Fig. 2. *Growth of two hypothetical micro-organisms as a function of substrate concentration. Low substrate concentrations are characteristic of soil, high concentrations are typical of laboratory media (see text for explanation)*

In Fig 2, the growth rates of two hypothetical organisms A and B are plotted as a function of substrate concentration. Clearly, B grows faster than A in low nutrient concentrations and would therefore be more competitive than A in soil, but not under laboratory culture. One reason for such changes might be that different metabolic processes occur under variable nutrient conditions. Organisms grown in nutrient-rich conditions form large amounts of secondary and intermediary metabolites, which are secreted from the cells, or storage products that are retained within the cell[185]. However, in growth-limiting conditions there may be a virtual 100 per cent efficiency in the conversion of substrate carbon into cell structural materials by some wild-type organisms[366].

Although there are no reliable methods for assessing the growth-rate of micro-organisms in natural soil (see Chapter 3), some idea of maximum growth rates possible can be obtained from data on energy input, population size and metabolic activity. Babiuk and Paul[577] suggested that the energy input of a grassland prairie soil was sufficient to allow bacteria to divide only a few times a year. Gray and Williams[578] made similar calculations for a woodland soil and showed that even if all the energy was utilised by bacteria (a minor component of the soil population in terms of biomass), they could only divide once every sixteen hours. Calculations of the generation time of bacteria in agricultural soils based on rates of carbon dioxide output also suggest that bacteria could divide about once a day. Since the energy-yielding materials have to support large populations of fungi and soil animals as well as bacteria, these estimates of generation time are probably too low.

Of course, organisms in the soil are not always present in near growth-limiting conditions, for on the surface of living roots and organic particles, nutrients may be abundant, at least for a short period of time. However, because of changes in the environment or because key nutrients are exhausted, organisms must be able to respond to the onset of starvation. When bacteria are starved, their proteins break down at an accelerated rate and provide a larger amino acid pool within the cells[345]. This pool may then be used to resynthesise new enzyme proteins, even though no growth (net protein synthesis) takes place[305]. This could explain how enzyme induction takes place, even in starved cells presented with a new and previously unavailable nutrient. Resynthesis of enzymes requires energy and it has been suggested[562] that it is released from reserve materials in the cell, e.g. glycogen, poly β-hydroxybutyric acid, sugar alcohols or starch. Other changes which accompany starvation include the induction of sporulation in fungi[396], blue-green algae[572] and bacteria[304], the formation of starvation-resistant mutants[191] and the formation of melanin pigments in some fungi, e.g. *Aspergillus nidulans* (see p. 104). Since the environment plays such a key role in determining the status of micro-organisms within the soil, both structurally and physiologically, its nature and effects on the soil population will be considered in the next chapter.

Viruses

Viruses, unlike the other organisms that we have considered so far, are non-cellular and can grow and multiply only inside living cells of animals, insects, plants, fungi, blue-green algae, bacteria and actinomycetes. For this reason, they must be dealt with separately. Chemically, viruses consist of protein and nucleic acid. Plant viruses contain only RNA, bacterial viruses (bacteriophages) usually only DNA, and animal viruses either DNA or RNA. The protein is usually in the form of spherical or spheroidal sub-units, packed in a symmetrical manner to form polyhedral or rod-shaped structures. In the case of bacterial viruses, a tail is joined to the polyhedral head which, together with the tail fibres, serves to attach the organism to the host bacterium, and acts as a channel through which the DNA may be passed and injected into the bacterium.

Virulent bacteriophages are common in soil and can lyse and kill a wide range of bacteria when in culture. Some of these phages are specific to certain bacterial strains while others attack a broad range of strains, species or even genera of bacteria. The role of the bacteriophage in controlling bacterial populations in soil is discussed in Chapter 9.

Most plant viruses are unable to survive in the soil for long periods of time, presumably because they cannot multiply outside host cells and because they would be degraded by microbial enzymes. A few viruses apparently do survive for a short time in decaying vegetation, e.g. wheat rosette virus, tobacco necrosis virus and tobacco mosaic virus. These sources provide viruses, which enter new plants through wounds in roots and root hairs. Wheat rosette virus is interesting because it will give rise to infection only if infected roots remain in soil, but not if infected leaves or sap remain. Soil free from higher plants can remain infective for up to five years. This evidence, coupled with the difficulty of successfully producing mechanical infections, suggests that the virus may be carried by a soil-borne vector and that it does not really survive in the free-living condition. Indeed, Teakle[505] has shown that tobacco necrosis virus may be transmitted by *Olpidium brassicae*, a soil fungus belonging to the Chytridiales. Whether the virus is transmitted inside the fungus cells or merely attached to the outside is not known, although lettuce big-vein virus does seem to occur inside fungal spores. Another group of plant viruses, causing diseases such as raspberry leaf curl and peach yellow bud, appear to be transmitted by nematodes, e.g. *Xiphinema* spp. Generally, the incidence of these diseases parallels the distribution of the nematode vectors, so that killing the vectors will eliminate the disease. It is not clear whether any of these vectors are harmed by the viruses that they harbour.

Viruses that attack animals, especially insects in the Lepidoptera, Hymenoptera and Diptera, are well known, but there is practically nothing known about their distribution in the soil. There have been virtually no reports of viruses killing some of the more typical soil insects, e.g. the Hemiptera, Orthoptera and Coleoptera, but when insects are attacked it is principally the larval stages that are affected.

2

Soil as an Environment for Micro-organisms

The initial stage in the formation of soil is the weathering of rocks, resulting in the production of mineral particles. The gradual colonisation of such material by plants and animals causes the build up of organic matter, leading to the development of a mature soil. Eventually soils reach an equilibrium with the climate and vegetation, although the balance is by no means static. Many soil properties, such as texture and pH, may be measured on a macro-scale and the results used to characterise soil types covering thousands of square miles. On the other hand, it is often necessary for the microbiologist to consider environments of microscopic dimensions, commensurate in size with the microbes themselves. Some factors may even vary within such micro-environments and recently the importance of 'molecular environments' located at the surfaces of soil particles and microbial cells, where ion concentrations are sufficiently different from those in surrounding solutions to influence microbial activity, has been demonstrated[302]. Realisation of the importance of micro-variations in soil environmental factors is growing, but practical application of such concepts is limited by the lack of sufficiently sensitive techniques.

The various components of the soil environment are constantly changing and a fluctuation in one results in fluctuation of others. Thus an alteration in moisture content influences the aeration, temperature and reaction of a soil. Because of this, it is rarely possible to relate accurately the behaviour of soil microbes to any one environmental factor. Two main approaches have been made to this problem. Firstly, the soil in its natural state has been studied and measurements of one or more factors taken. Attempts have then been made to relate these to the behaviour of the micro-organisms, so that the presence or absence of a particular organism or group of organisms has sometimes been tentatively related to variation of one factor. The effects which are easiest to discern are the extreme ones, such as the limitation of microbial activity and viability by shortage of moisture. Secondly, the behaviour of soil organisms in artificial systems has been investigated where at least some of the environmental factors can be controlled (see Chapter 3).

In this chapter, the general characteristics of the soil environment will be outlined, and application of the approaches mentioned above will be considered. Since soil may be regarded as a three-phase system, composed of solids, liquids and gases dispersed to form a heterogeneous matrix, it is convenient to discuss each phase separately.

Solid Components

(a) *Mineral Materials*

As rocks are weathered, particles of different size and chemical composition are produced. Soils therefore consist of a mixture of different sized mineral particles, the proportions of which determine the soil texture. The larger particles (50 μm diameter) are mainly fragments of the parent rock material and have a low surface area per unit mass (specific surface). Silt particles (2-50 μm diameter) are also composed of primary minerals, e.g. quartz, but have a greater specific surface, whilst the clay particles, the smallest of all (<2 μm diameter) are composed of secondary minerals, e.g. kaolinite, montmorillonite and illite, and have a very large specific surface. By way of illustration, if a cubic-shaped soil particle with 1 mm sides was split into cubes with the dimensions of clay particles (*c.* 0·5 μm), the increase in surface area would be from 6 mm^2 to 12,000 mm^2.

Clay particles are extremely important components of the soil environment, and greatly influence the physical and chemical properties of soil. Clay minerals have a crystalline structure composed of several plate-like layers (Fig. 3) and in some, e.g. montmorillonite, these plates may move apart in the presence of water to expose their internal surfaces. This increases the specific surface of the particles still more, so that while a two-layered, non-expanding lattice such as kaolinite has a specific surface of about 50 m^2/g of clay, montmorillonite with a three-layered, expanding lattice has a specific surface of 800 m^2/g.

The overall electrical charge on clay particles is negative. This is due partially to atom replacements which can occur within the clay without disturbing lattice structure (isomorphous substitution); replacement of tetravalent silicon by trivalent aluminium, for example, leaves one negative charge from an oxygen or hydroxyl group unsatisfied. Another contribution to the overall negative charge comes from the terminal atoms on the edges of the lattice. The presence of these negatively charged particles has a considerable effect on the distribution of ions in the soil solution. Cations such as H^+, K^+, Na^+, Ca^{++} and Mg^{++} are attracted to the particle surfaces and form a layer of positive charges around them known as the *electric double layer*. Usually, this

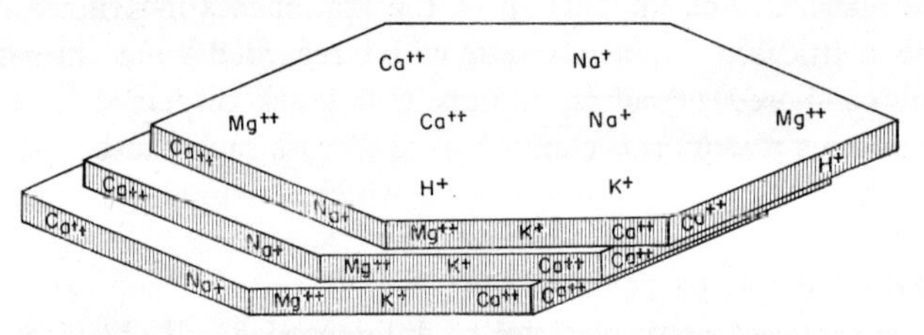

Fig. 3. *Diagrammatic representation of the plate structure of clay minerals and exchangeable bases on exposed surfaces and edges.* (*Chapman*[77])

(*Reproduced by permission of H. D. Chapman and the Regents of the University of California*)

layer is diffuse because the kinetic energy of the ions can move them into the surrounding solution against the attractive forces of the clay particles. This is especially marked with low valency cations which are less strongly held by the particles and which constitute the exchangeable cations of a soil. Low salt concentrations in the surrounding solution also promote the formation of a more diffuse double layer. In consequence, a gradient of cation concentration exists between the clay particle surfaces and the soil solution outside the influence of the double layer. Anions in the soil solution, e.g. Cl^-, NO_3^- and SO_4^{--}, are repelled from the clay particles and form a concentration gradient in the reverse direction. The extent of the double layer varies from 2 to 20 nm, but it has been calculated that as much as 0·5 ml of the soil solution per gram of soil would be influenced by clay particles with a specific surface of 100 m^2 g and a double layer of 5 nm, and would therefore differ in composition from the rest of the soil water[135]. When little water is present, it is possible that all ions in the soil solution are within the zone of influence of the particles. These micro-scale variations in the soil environment influence microbial activity in a number of ways, especially in relation to hydrogen ion concentration (see p. 42).

Certain clay minerals present in soil have recently been shown to affect the activity of soil micro-organisms. Montmorillonite, when present in low concentrations, reduces the rate of fungal respiration in laboratory cultures, while stimulating that of bacteria in either low or high concentration[492, 493]. In subsequent work[488], it was suggested that this stimulation is due to the physico-chemical properties of montmorillonite. This mineral, with its expanding lattice and high specific surface, has a considerable cation exchange capacity and can mediate the pH of systems by the replacement of hydrogen ions, produced by microbial metabolism, with basic cations from its exchange complex. Thus it could help to preserve a pH conducive to bacterial activity. The inhibitory effect on fungi was thought to be caused by the high viscosity of montmorillonite which impaired the diffusion of oxygen[494].

Many mineral particles become bound together in soil crumbs; the size of these aggregates can be influenced by the microbes both through the binding properties of fungal and actinomycete mycelium and by polysaccharides produced by bacteria (see Chapter 6).

(*b*) *Organic Material*

Organic materials from plants and animals both above and below ground are constantly supplied to soil where they are decomposed by its micro-organisms. Almost any naturally occurring chemical may be present in soil at one time or another, albeit for only a short period, and man's activities have added many new chemicals to soil in the form of artificial fertilisers, herbicides, pesticides and industrial wastes (see also Chapter 5).

Organic matter which is not completely decomposed eventually contributes to the formation of amorphous humic materials which are an important aspect of the soil environment. Humus is not a defined chemical entity but a complex mixture of many substances. Within this complex there is a small fraction of water-soluble organic substances (e.g. amino acids, sugars) but by far the largest part consists of insoluble, dark coloured material. This may be divided into three fractions:

(*i*) Humic acid, extractable with alkali but precipitated with acid;
(*ii*) Fulvic acid, extractable with alkali and soluble in acid;
(*iii*) Humin, not extractable with alkali.

Humic acid is extremely resistant to decomposition and is probably formed by the polymerisation of compounds derived from lignins, proteins, plant and animal metabolites. It is believed to consist of an aromatic core which has an aliphatic periphery. However, the exact composition of humic acid is not fixed and varies from soil to soil and even from one part of the same soil to another[61] (see Chapter 5).

Fulvic acid is not an homogeneous substance and contains carbohydrate and protein material in addition to fractions like those found in humic acid. Humin is also heterogeneous, containing unchanged plant residues and also probably humic acid in association with clay colloids which prevents its extraction.

Humic material has colloidal properties and its capacity to hold and exchange basic cations can exceed that of clay minerals in some soils. It forms associations with mineral colloids and may be absorbed on to the surfaces of larger mineral grains. Alternatively, humic materials may move through the soil under certain conditions, resulting in an uneven distribution of organic matter through the soil profile. This is particularly marked in podzols and some other soil types (see p. 48).

Liquid Components

(a) General Characteristics of Soil Water

Soil water is subject to extreme fluctuations and these have important effects on the soil microbial populations and plant roots. Soil water with its many solutes, including minerals, organic substances and gases, forms the liquid nutrient medium for micro-organisms, and as the amount of water in the soil varies, so the concentration of its solutes changes.

On the basis of its distribution and behaviour, four types of soil water may be recognised:

(*i*) Gravitational, which drains out of soil under the influence of gravity and is generally available to soil organisms; also plays a major role in the transport of materials;

(*ii*) Capillary, which is held in the pore spaces and has a similar role;

(*iii*) Osmotic, which is held around clay particles and humus and is less available to microbes and roots;

(*iv*) Hygroscopic, which is very strongly adsorbed by particles, forming a very thin film around them, and is the least available form of soil water.

Osmotic and hygroscopic states of water bound tightly around particles are difficult to distinguish from water held in very fine pores between such particles. This is particularly true of badly structured clay soils in which most pores are less than 1 μm in size.

The state in which water occurs may be roughly related to the quantity of water in a soil. If soil is saturated with water, the excess will drain out under gravity. The amount remaining after removal of all gravitational water is termed the '*moisture holding capacity*' of soil.

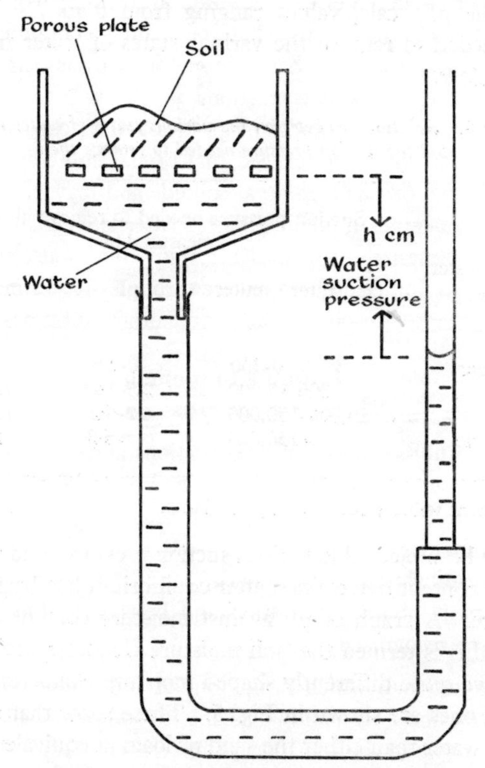

Fig. 4. *A water manometer illustrating the principles of water suction pressure. The difference between the water levels in the manometer (h cm.) is a measure of the water suction pressure on the soil*

As the soil becomes drier, the pore spaces become progressively filled with air, until all capillary water is removed leaving only the tightly bound osmotic and hygroscopic water.

It is evident that as soil dries, it becomes increasingly difficult to remove the residual water. It is possible to measure, by a variety of methods, the suction pressure needed to bring soil to a particular moisture level. This pressure is normally measured in centimetres of water suction and a simple apparatus showing the principle of this is shown in Fig. 4. Because of the wide range of suction pressures involved (0 to 10,000,000 cm water suction), it is more convenient to express the values as $\log_{10}$. The logarithmic conversions form what is known as the pF scale, values ranging from 0 to 7. The suction pressures needed to remove the various states of water from soil are given in Table 9.

Table 9. *Relationship between the various states of water in soil and the suction pressure needed to remove them*

State of water	Suction pressure needed to remove all water measured as		
	cm. water suction	pF	atmospheres
Gravitational	0-300	0-2·5	0-0·3
Capillary	300-15,000	2·5-4·2*	0·3-15
Osmotic	15,000-150,000	4·2-5·2	15-150
Hygroscopic	>150,000	>5·2	>150

* Permanent wilting point of higher plants.

A soil may be subjected to various suction pressures and the percentage moisture content determined after equilibrium has been reached at each pressure. A graph of pF against moisture content can then be plotted and this is termed the 'soil moisture characteristic'. Different soils may have quite differently shaped moisture characteristic curves; some typical ones are shown in Fig. 5. These show that the clay soil retains more water than either the sand or loam at equivalent pF levels. In the sandy soil, with few colloidal particles, most of the water will be held in relatively wide pores (> 1·0 μm), whereas in the clay a greater proportion will be held in fine pores (< 1·0 μm) or present in the osmotic and hygroscopic state. The loam soil is intermediate in its behaviour since it contains a mixture of both clay and sand particles. Clearly, the energy which a microbe or plant root must expend to obtain a given amount of water at any one moisture level is different in each of these soils. Thus, if each soil contained 10 per cent water, a micro-organism would have to exert a suction pressure of at least pF 1·5 in the sand, 4·0 in the loam and 5·2 in the clay to obtain the water (see below).

Gross measurements of the water content or pF of a soil generally conceal the fact that water is unevenly distributed within the soil.

While some parts of the soil surface are sheltered from rain, others receive increased supplies of water through run-off from tree trunks and canopies. Once rain reaches the soil, it may also pass through it

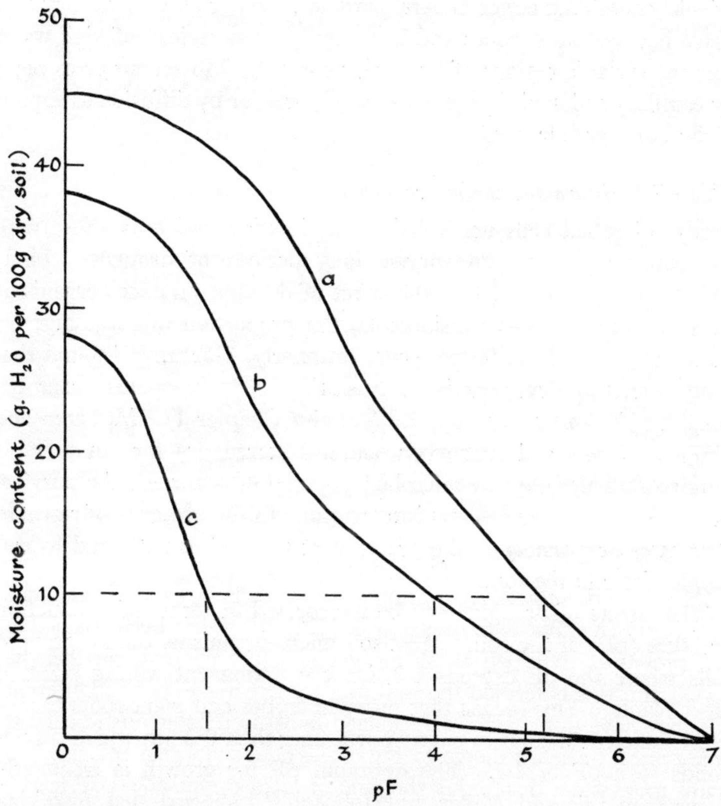

Fig. 5. *Representative moisture characteristic curves.* (a) *Drying boundary curve of a clay soil.* (b) *drying boundary curve of a loam soil.* (c) *drying boundary curve of a sandy soil.* (*Griffin*[170])

(*Reproduced in modified form by permission of D. M. Griffin and the Cambridge Philosophical Society*)

unevenly, as clearly demonstrated by Reynolds[394]. Fluorescent powders, which he placed on the soil surface, dissolved in the rainwater and so traced the passage of water in the soil. Greater downward percolation was observed near tree trunks, while most lateral movement occurred in the humus layer. Low moisture contents are often found around roots because of their efficient uptake; rainwater percolates more easily through these drier zones and the cracks and channels made by roots. Other experiments have shown that within a soil, high moisture contents are generally associated with high concentrations of colloidal material, whilst low moisture contents are characteristic of soils containing larger mineral particles. Water movement may take place between such zones and is generally from regions of high water content towards regions of low water content. Movement takes place by capillary action along the network of pores, or by diffusion as vapour if the pores are air filled.

(*b*) *Soil Moisture and the Microflora*

Many microbial cells are killed by desiccation and only those with resistant propagules can survive long periods of drought. Thus, Meiklejohn[318] showed that the effect of drought on micro-organisms in a Kenyan soil was considerable, the proportion of actinomycetes increasing from 30 to 90 per cent. Similarly, Warcup[541] found that fungi survived dry periods in Australian soils as spores, sclerotia, rhizomorphs and resting hyphae (see also Chapter 1). McLaren and Skujins[302] reported that when soil was stored dry for ten years, a hundredfold decrease in microbial population occurred; the survival of the remaining micro-organisms was due to the presence of resistant structures or possibly to the presence of residual osmotic and hygroscopic water in the soil.

The effects of drought may be interpreted in terms of the suction pressure (pF) of the soil. Most soil micro-organisms cannot grow in soils where the pF exceeds 4·2, i.e. the permanent wilting point of most plants. This means that many microbes and plant roots cannot exert sufficient suction to empty pores finer than 0·2 μm which corresponds to a pF of 4·2. The optimum pF for growth is frequently lower than this and Miller and Johnson[323] showed that nitrifying bacteria grew best when the pF was less than 3·0, while Rahn[385] found that the optimum thickness of the water film for the growth of *Bacillus cereus mycoides* was between 20 and 40 μm, which corresponds to a pF of 2·0.

One of the best methods for determining the effect of pF on the growth of micro-organisms is to grow them in artificial soils in which both water availability and pore size can be controlled. This approach has been used by Griffin[167-170] with considerable success. He has shown that the responses to different moisture regimes are not the same for all fungi. While a few forms, e.g. *Aspergillus restrictus*, are still able to grow to some extent at pF 5·6, most stop growing at pF values in excess of 4·2. As artificial soils are made wetter and the pF values drop, other effects are observed. For instance, when the pores are filled with water, *Curvularia ramosa* starts to grow abnormally and at a slower rate, although *Pythium ultimum*—the damping-off fungus—is unaffected. Shameemullah[442] attempted to obtain similar data for fungi growing in natural soils. He found that some organisms are very sensitive to increases in pF, e.g. *Mortierella* spp. and bacteria while others like *Penicillium* spp. and the actinomycetes survive high suction pressures.

Similar experiments on soil algae have not yet been performed, partly because it is difficult to isolate representative members of the terrestrial microflora. However, it is only reasonable to expect to find great differences among them in respect of their response to water since algae are to be found not only in bodies of water but also as important colonisers of desert soils.

From the above experiments on the effect of water on soil organisms, we might conclude that remoistening dried soil would result in the restoration of microbial activity. However, the level of activity reached often exceeds that which existed before drying. It has been suggested that this is due to the effect of drying causing an increase in solubility of nutrients which become available upon remoistening[175] although Greenwood[161] has pointed out that a re-orientation of organic matter might take place during drying, caused by movement of particles and disruption of pores and making some previously inaccessible substrates available to microbes. Drying of soil also kills some micro-organisms which then provide a source of nutrients for the survivors.

Gaseous Components

(*a*) *The Pore Space System*

Between the soil particles and crumbs are the pores, which form a continuous network filled with liquids and gases. Most soils have a pore space comprising about 50 per cent of their total volume. The precise proportion of pore space in any soil is affected by vegetation and treatment. Gadgil[138] showed that the upper layers of a soil under permanent pasture had a larger amount of pore space than did a young ley planted with grass. This was attributed to the action of roots and micro-organisms on crumb structure (see p. 153).

The size of the pores may become too small to allow normal growth and sporulation. As one might expect, fungi, with large vegetative and sporing bodies, are more susceptible than the smaller bacteria or actinomycetes. Griffin[167, 168] found that pore sizes in the range 2·7 to 100 μm had no marked effect on the rate of hyphal growth of *Pythium ultimum* or *Curvularia ramosa*. However, sexual reproduction in *Pythium* was greatly reduced when the pore diameter reached 15 μm (the diameter of an oogonium being about 20 μm). Similarly, sporulation by *Curvularia* was diminished when the pore radius fell below 23 μm (the length of a spore being about 27 μm).

The relative proportions of gas to liquid in pore spaces are constantly changing. A sand that is dried almost to wilting point will have almost all of its pore space filled with air, but a poorly structured clay will have no air space when wet. The composition of soil air differs from that of the atmosphere above. It is usually saturated with water vapour and contains more carbon dioxide but less oxygen; in addition, gases formed by microbial activity, such as ammonia and methane, may be present together with other volatile products. Some of these gases are very soluble and also occur in the soil water (e.g. ammonia, carbon dioxide). Consequently there is an exchange of gases between the soil and the atmosphere through air-filled pore spaces, with oxygen diffusing into soil and carbon dioxide moving out. This tends to reduce carbon dioxide concentrations in the upper parts of a soil but has less effect on the lower regions and hence concentrations increase with depth in most soils. The composition of the air also varies from point to point within the soil; in regions of intense microbial activity, especially near plant roots or animals, concentrations of gases will be radically altered. The constant movement of gases within the pores is again from regions of high concentration towards those of lower concentration. This movement is much more rapid if the pores are not filled with water, as diffusion of gases in liquids is much slower. The rate of diffusion of oxygen through air is about 10,000 times more rapid than it is through water, so that the aeration of soil is markedly influenced by its moisture content.

(*b*) *Soil Aeration and the Microflora*

Microbes vary considerably in their responses to changes in the concentrations of gases in the soil atmosphere. Some bacteria are strictly anaerobic and are unable to grow in the presence of oxygen (e.g. *Clostridium botulinum*) while others are strictly aerobic (e.g. *Pseudomonas fluorescens* and most actinomycetes). Fungi are also strict aerobes, and as in other groups, the quantitative relationship between growth and oxygen supply varies in different species. Some bacteria appear to grow well only in very low oxygen tensions and are termed microaerophiles. Casida[70] found such forms in considerable numbers in soil and these have recently been identified as *Streptococcus*-like organisms[145] and actinomycetes.

The balance between aerobic and anaerobic conditions in soil seems to be a fine one. Greenwood[161] estimated that the change from aerobic to anaerobic metabolism takes place when the oxygen concentration becomes less than 3×10^{-6} M. On this basis, Greenwood and Berry[164] suggested that it was likely that water-saturated soil crumbs, about 3 mm or more in radius, would have no oxygen at their centres, even if surrounded by air. Thus it is probable that anaerobic micro-environments occur in most soils, allowing the growth of anaerobic bacteria, including those which can fix nitrogen (e.g. *Clostridium pasteurianum*). Anaerobic conditions often lead to the accumulation of sulphides and ferrous iron, with a decrease of nitrate and available phosphates. Decomposition of substrates by microbes under anaerobic conditions may result in the formation of incompletely oxidised end-products. When cellulose is decomposed by bacteria in aerobic conditions, the products are carbon dioxide and water, but in anaerobic conditions, oxidation is incomplete and organic acids, such as acetic and formic acids, accumulate. The concentration of oxygen affects the oxidation-reduction potential of soil, but it is not known whether this in itself is a determining factor in microbial distribution or simply a reflection of the oxygen status of the environment[7].

The concentration of carbon dioxide in the soil atmosphere is an important factor in itself, and can have three main effects on soil micro-organisms[7]. It influences the pH in micro-habitats, provides a source of carbon for autotrophic microbes and has a differential inhibitory effect on the heterotrophic microflora. Burges[58] considered that high concentrations of carbon dioxide in badly aerated parts of soil could be more important than low levels of oxygen in limiting fungal activity. Stotzky, Goos and Timonin[490] incubated cultures of soil microbes under carbon dioxide and found that growth of 99·9 per cent of the microflora was inhibited. However, later work by Stotzky and Goos[489] showed that soil microbes were, in general, tolerant of high carbon dioxide and low oxygen concentrations and they suggested that extreme fluctuations in the soil atmosphere did not inhibit microbial development to the same extent as these extremes inhibited more specialised populations in foodstuffs. More fungi developed under conditions of low oxygen and high carbon dioxide concentration than either bacteria or actinomycetes. Some fungi, for example *Fusarium* species, developed particularly well under these conditions.

The behaviour of specific fungi in controlled atmospheres, containing various proportions of nitrogen, oxygen and carbon dioxide, known to occur in air-filled pores in soil, has been studied by Griffin[171] (Table 10). Little effect on fungal development occurred until extreme

Table 10. *Fungi isolated from ribwort stems in contact with a saline soil held in various gas mixtures, modified from Griffin*[171]

	N_2 78·1	O_2 20.9 (air)	CO_2 0.03	N_2 81·0	O_2 5·1	CO_2 13·9	N_2 81·5	O_2 0·93	CO_2 17·6
Curvularia geniculata		3			1				1
Fusarium sect. *Gibbosum*		4			4				4
Pythium debaryanum		1			3				4
Pythium periplocum		4			1				0
Rhizoctonia sp.		2			1				0
Trichoderma köningi		4			4				4

0—absent
1—very rare
2—uncommon (on about 25 per cent of stems)
3—common (on 25-50 per cent of stems)
4—abundant (on more than 50 per cent of stems)

gas mixtures were applied, i.e. those which occur in water-filled pores, when marked inhibition of growth was observed. It was concluded, therefore, that poor aeration has a marked inhibitory effect on soil fungi only when the pores are filled with water and rates of gaseous diffusion are low. *Fusarium* spp. were again found among those tolerating high carbon dioxide concentrations, together with *Trichoderma köningi*, and *Pythium debaryanum*, the latter becoming more abundant as carbon dioxide concentrations increased.

Soil Reaction and the Microflora

The pH of a soil is determined by a number of factors, including the concentrations of salts and carbon dioxide in the soil solution and the exchangeable cations present. As these factors fluctuate both in space and time, so does the pH of a soil, usually within the limits of one to two units. As water moves through a soil, there is a tendency for bases to be leached out and replaced by hydrogen ions; thus constant leaching leads to the formation of an acid soil. These changes can affect the chemical properties of soil, altering the solubility of certain substances. Under acid conditions, many substances become more soluble and some, such as calcium and phosphorous compounds, may be leached away, resulting in nutrient deficiencies. Increases in the solubility of aluminium, iron, nickel and other compounds may result in their levels in the soil solution becoming toxic to microbes and plants[213].

The pH of microbial cytoplasm approximates to neutral and the majority of soil micro-organisms grow best at pH values near to 7. There are a number of exceptions to this, the most spectacular being bacteria belonging to the genus *Thiobacillus*, which are acidophilic and will grow even at pH 0·6. Generally, however, soil bacteria and actinomycetes are less tolerant of acid conditions than are fungi. The critical pH level for most bacteria and actinomycetes is around 5 and below this many cease to grow. There is evidence for the existence of acid-tolerant *Streptomyces* species. Some strains isolated from an acidic podzol were found to be more tolerant of acidic conditions in culture than those from a less acidic soil[95]. Many species of fungi can grow in acidic conditions and in soils such as podsols, where pH may be as low as 3, fungi are the dominant members of the microflora. Considerable attention has been paid to such soils by mycologists, probably because of the attractions of obtaining isolation plates free from troublesome bacteria. The fungus flora of these soils is characterised by species such as *Mucor ramannianus*, *Mortierella parvispora*, and *Trichoderma viride*[565]. However, the pH preferences of fungi are various and species in the same genus may have quite different optima; *Mortierella parvispora* is associated with acid soils, while *M. alpina* is found in alkaline conditions. Similarly, Dick[110] found that species of *Saprolegnia* in soil had widely different pH ranges and optima.

The occurrence of a microbe in a soil with a certain gross pH does not mean that it is living and carrying out all its activities at that pH. In micro-environments the reaction is not always the same as that of the macro-environment, the activities of soil organisms and plant roots together with the chemico-physical properties of soil particles contributing to localised changes of pH in either direction. McLaren[298] pointed out that the negatively charged colloidal clay particles attracted hydrogen ions and hence the concentration of these would be higher around such particles. Enzymes released by micro-organisms can also be adsorbed on to the particles and hence be subjected to a lower pH than that in the ambient solution. This was demonstrated by McLaren and Skujins[301] who found that, in liquid culture, the pH needed for the half-maximum rate of nitrification by *Nitrobacter agilis* was about 0·5 unit higher in the presence of soil particles than in a solution without soil. This rise in gross pH was necessary to counteract the drop in pH around the colloidal particles, where hydrogen ions accumulated. In this case, availability of nitrite for conversion to nitrate was also altered at particle surfaces, concentrations being lower because the negatively charged nitrite ions were repelled. The rate of nitrification was consequently lower in the presence of soil particles.

The pH at particle surfaces is not always lower than the gross pH of a soil. When amino acids, urea or chitin are added to an acid, sandy soil (pH 4·0), decomposition and subsequent release of ammonia can lead to an increase of soil pH to values of 6·0-7·0. Microscopic examination of such soil, flooded with a pH indicator, has shown that increases in pH may occur around particles of organic matter where ammonia is held (Mayfield, unpublished). Thus in natural soil, microbial activities can sometimes counteract local increases in hydrogen ion concentration.

The situation may be further complicated by charge effects at the surfaces of the microbes themselves. Weiss[556] compared the activity of penicillinase from *Bacillus subtilis* in cell-bound and free state over a range of pH; the enzyme bound to the outside of cells behaved as if it was in an environment of lower pH than that of the bulk phase.

Soil Temperature and the Microflora

There are several factors which determine the amount of solar energy absorbed by soil, including the direction and degree to which the land surface slopes, the soil colour and the density of the vegetation cover. Lighter coloured soils absorb heat less efficiently than dark ones and in the northern hemisphere, southward-sloping land receives most solar energy per unit area. Vegetation can further reduce the amount of heat reaching soil. This is most marked with plants, such as conifers, which not only intercept radiant energy with their leaves but also form a thick layer of litter which insulates the soil. Similarly, the deeper parts of soil are buffered by the upper layers and experience less temperature fluctuation. Wilkins and Harris[561] found that average monthly temperatures on the surface of a forest soil varied from 2° to 19°C, whereas at only 7·5 cm depth the range was reduced to 4-14°C.

A close association exists between the moisture content of soil and its heat-absorbing capacity. Whereas 1 calorie is needed to raise the temperature of 1 gram of water through 10°C, only 0·2 calories are required to raise the temperature of 1 gram of typical soil solids to the same level[386]. However, the thermal conductivity of a wet soil is higher, so that heat is conducted downwards more efficiently. Thus although the surface temperature of a dry soil is often higher than a wet one, the temperatures a few inches below the surface may be identical. As moist soil absorbs solar energy at its surface, the hydrogen bonds between water molecules are broken and water evaporates from the soil. This produces a water deficit at the surface and so water moves upwards from the deeper layers. At the same time, the high surface temperature causes a downward flow of heat. As the temperature rises, so does the diffusion rate of gases, resulting in a complex interaction between temperature, moisture and aeration. However, the increased rate of supply of gases caused by increased diffusion rates are probably largely offset by the extra demand brought about by higher rates of respiration.

Micro-organisms can be placed into one of three groups on the basis of their temperature requirements, but these groups overlap considerably and there is no general agreement on their limits.

(*i*) Psychrophiles can grow at low temperatures (below 5°C) but may still have a high optimum growth temperature, similar to mesophiles[122];

(*ii*) Mesophiles do not grow at low temperatures or at high temperatures above about 40°C. Their optimum growth is usually between 25° and 37°C;

(*iii*) Thermophiles can grow at high temperatures (45° and 75°C) with an optimum between 55° and 65°C and little growth below 40°C.

The vast majority of soil microbes are isolated by incubation at temperatures suitable for optimum mesophile growth and most are mesophilic in their response to temperature ranges in culture. However, ambient soil temperatures in many cases are unlikely to reach levels suitable for maximum mesophilic development. An even greater dilemma is faced when thermophilic microbes are isolated from such soils or when mesophilic microbes are detected in extremely cold soil. It seems unlikely that so many micro-organisms are adapted to grow best at temperatures never occurring in their natural environments.

A possible explanation is that temperatures in micro-environments, where there is concentrated chemical and biological activity, can exceed those obtained for the bulk of the soil.

Soil temperature has an obvious effect on the general metabolic activity of the microflora but does not often have a direct lethal action. The extremes of temperature to which soil microbes are subjected depends on their position in the profile and the climatic conditions above the soil. Those growing on or near the surface may be subjected to considerable changes in temperature during the course of a day, while microbes living in the lower parts of the soil may experience only small-scale changes in temperature throughout the year. In the tropics a soil may be at a high temperature all year, while in temperate regions maximum temperatures are lower and the annual range greater. Okafor[341] gave temperatures, taken at 1 foot depth, for a Nigerian soil and an English soil throughout the course of the year. In the tropical soil, the temperature was always between 28° and 30°C but in the temperate soil it ranged from 2° to 15°C. The organisms which colonised chitin buried in soil in these two temperature ranges were quite different; at 10°C fungi and bacteria were dominant, while at 20°C actinomycetes, nematodes and protozoa were important. Low temperatures also affect microbial activity, and while not necessarily killing soil microbes, at temperatures below freezing point many survive only in a state of low activity. Ross[403] studied the oxygen uptake of soils which had been stored at low temperatures and found that the potential activity of the micro-organisms had not been markedly changed.

Soil Profile Development and Soil Types

If a vertical section (or profile) of an uncultivated soil is examined, it is often possible to see several distinct horizontal bands, which are known as horizons. Adjacent horizons may have quite different physical and chemical properties and a factor such as pH may change markedly over a few centimetres. The development of soil profiles, beginning with the parent mineral materials, is a gradual process and hundreds of years may elapse before a mature profile is developed by the complex interactions of soil, climate, vegetation and the many environmental factors discussed earlier. The main processes discernable in profile development are the following:

(*i*) Deposition of organic matter on the soil surface and its incorporation into the soil;

(*ii*) Elluviation, or the movement of water down the profile, leaching soluble materials from the upper layers;

(*iii*) Illuviation, or the accumulation of materials, carried down from the upper layers, in the lower parts of the profile.

In some instances, plant litter falling on to the surface may accumulate and reach a depth of 10 cm or more. This happens particularly in temperate and cold climates on soils under coniferous trees and results in the formation of *mor* litter. Under deciduous trees, less litter accumulates and this type is known as *mull* (see Chapter 4). When a distinct, permanent accumulation of litter above the soil does occur, it is generally referred to as the A_0 horizon. Constant elluviation of the upper parts of the soil leads to a loss of soluble chemicals, particularly basic cations like sodium, potassium and calcium, which in turn causes a drop in pH in these regions. These elluviated horizons are referred to as the 'A' horizons. In acid conditions, other components such as iron and humic compounds may become mobilised and be moved to lower parts of the profile where some accumulate, forming the illuviated 'B' horizons. It is not clear why these materials lose their mobility and accumulate but the mobilisation of iron can be caused by extracts of low molecular weight from leaves and decaying organic matter[33, 109]. The parent material, from which the primary minerals of the soil were derived, underlies the other horizons and is known as the 'C' horizon.

Not all these horizons are necessarily represented in a soil and the presence and distribution of the various horizons enables us to recognise different soil types. Many types are recognised and many schemes for classifying them exist, but it is beyond the scope of this book to describe them. Instead, a few of the most widely distributed general types will be briefly described.

(*a*) *Brown Earths*

These develop in conditions of moderate to high rainfall over a basal material with low permeability, such as clay or loam. Hence the soil is subjected to moderate leaching, with the resulting formation of an acidic A horizon. The pH tends to increase gradually down the profile in which there are usually three horizons represented (Fig. 6a). The accumulation of litter is not marked and the border between the A and B horizons is not always clearly defined. Brown earths are typically found under deciduous forests and it is from this type of soil that most agricultural land in Britain is derived.

(b) Podzols

These are usually formed where the combination of high rainfall and a freely permeable sandy soil leads to heavy leaching of the profile, with the consequent formation of acidic conditions. The horizons are well developed and up to five or more may be present (Fig. 6b). The

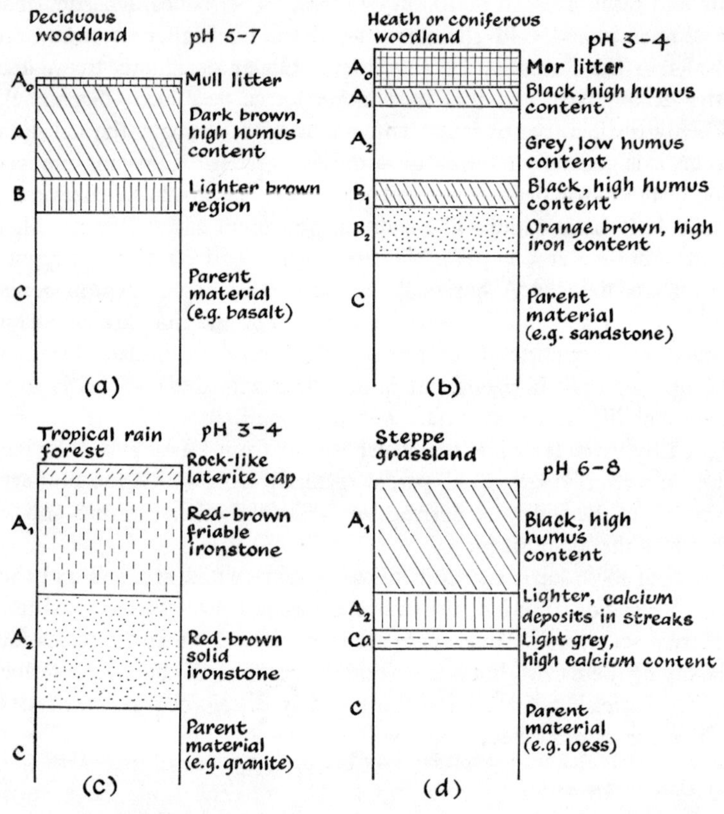

Fig. 6. *Diagrammatic representations of soil profiles.* (a) *Brown earth.* (b) *podzol.* (c) *lateritic soil.* (d) *chernozem*

A_0 horizon is well developed with considerable accumulation of litter and the elluviation of organic materials, together with iron and aluminium sesquioxides from the A horizons, leads to the formation of B_1 and B_2 horizons. Sometimes iron compounds accumulate in the B_2 horizon and cement together the particles, forming a hard 'iron pan'. Podzols are frequently found under coniferous and heath vegetation.

(c) Lateritic Soils

These soils are associated with conditions of high rainfall and extreme leaching in tropical climates. Under high temperature conditions, the soils are often red in colour, due to the formation of non-hydrated forms of iron and aluminium oxides (e.g. haematite and bauxite). In extreme cases, the upper layers are very rich in these oxides and the silica content is very low, dropping to 3 per cent or below, as a result of the desilification of the parent material. Little or no organic matter accumulates as its decomposition is extremely rapid, and complete breakdown to carbon dioxide and water, without formation of acidic intermediates, may occur (Fig. 6c). If surface soil layers are removed by erosion (which is encouraged by removal of forest vegetation by man), horizons of iron accumulation are exposed. These harden when subjected to drying and form an iron pan. Thus even in conditions of high rainfall the quality of the soil may be as low as that in desert regions. Laterisation is most widespread in soils of Africa, South America and South-east Asia.

(d) Chernozems

Chernozems (or black earths) form in a climate where there is sufficient moisture and suitable temperatures to promote weathering of minerals and humus formation, but not excessive leaching. They are typical of the grassy steppe and prairie regions with a low rainfall. The profile is quite characteristic, consisting of a black soil resting on its parent material which is usually loess and rich in bases. Sometimes A_1 and A_2 horizons can be recognised, the former being somewhat darker in colour, but no B horizons are present (Fig. 6d). The whole profile is saturated with calcium; free calcium carbonate may not be present in the A_1 horizon, but in the A_2 horizon, white streaks of calcium carbonate or sulphate may be seen, often associated with root channels. In addition to the supply of plant litter from above, these soils are permeated by grass roots which also supply organic matter.

Occasionally specific micro-organisms seem to have an association with a particular type of soil, although the reasons for these associations are not usually clear. One of the most clear-cut relationships is that of the fungus *Mucor ramannianus* with soils having an iron pan (B_2) horizon[441]. This seems to be an indicator species of these soils. However, such relationships are the exception rather than the rule, indicating the overriding importance of the micro-environment in controlling microbial distribution.

Seasonal Variations and the Soil Microflora

Most factors in the soil environment vary with climatic changes throughout the year. This is especially true of temperature, moisture content and nutrient supply. Qualitative and quantitative changes in soil populations have been related to seasonal variations in environmental factors, but such results cannot be accepted unless they are shown to be significantly different from spatial variations observed in the same populations[563]. Often clear seasonal patterns do not emerge and there are two main reasons for this. Firstly, much of the soil is buffered against large-scale changes, and secondly, the methods used to detect changes in microbial populations are not sensitive to small-scale fluctuations. In soils subjected to extreme climatic conditions, such as long periods of drought which occur regularly each year, clear seasonal changes in the microflora have been detected[541, 318]. In less extreme climates, one of the most regularly fluctuating factors is that of nutrient supply, flushes of nutrient occurring after leaf fall. Nicholas, Parkinson and Burges[330] found a relationship between the amount of fungus mycelium in a podzol and the fall of pine needles, maximum quantities occurring in autumn and winter; Ross[404] measured the mean annual oxygen uptake of soil and showed that it was related to organic matter, i.e. nutrient supply, rather than to water content or rainfall.

The unravelling of the effects of environmental factors on organisms is difficult in any habitat. In soil, the problems of measuring factors with accuracy on a micro-scale must be added to the difficulties presented by factor interaction. The best hope for the future probably lies with the use of artificial systems approximating to soil, in which factors can be controlled and accurately measured. Information on the responses of microbes in such systems could then be used to make predictions which might be tested in natural soil.

3

Methods of Studying the Ecology of Soil Micro-organisms

Much of the science of microbiology is concerned directly or indirectly with assessment of biochemical reactions brought about by micro-organisms. To study these reactions, methods have been evolved in which organisms are grown in pure culture under conditions which encourage maximum yields. Factors which affect these yields have usually been studied singly in homogeneous and controlled conditions. The development of soil microbiology has been much influenced by this work and many of its methods have been applied to experiments with organisms isolated from soil However, the application of conclusions drawn from such experiments to organisms growing in the soil has not always been successful, for the problems are far more complex. As we saw in the previous chapter, soil is a fluctuating, heterogeneous system of gases, liquids and solids in which mixtures of organisms are growing on very small amounts of nutrients. Consequently, although many of the methods may be used, the interpretation of the data obtained with them is difficult for it is never certain that the conditions reproduced in the laboratory exist in micro-environments within the soil. This inability to define clearly the limits of particular environmental conditions also results in major sampling problems and consequent difficulties in treating the data statistically.

Many of the techniques used to study pure cultures under laboratory conditions are fully described in the literature and will not be discussed here. In this chapter, we shall concentrate upon those methods which give information on the condition of the organisms in the soil. The techniques used for this can be grouped conveniently according to the purposes that they serve, the most important ones being as follows:

(*i*) Determination of the form and arrangement of micro-organisms in soil;
(*ii*) Isolation and characterisation of micro-organisms;
(*iii*) Detection of microbial activity in soil;
(*iv*) Determination of microbial biomass.

Many techniques are available in each of these categories. None of them is perfect and it is generally useful to apply a number of techniques, if possible based on different principles, to try to build up a complete and accurate picture. The usefulness of any one technique is often limited by the biological properties of the organisms under study, so that a method satisfactory for one group of organisms may provide unreliable data for another group. In the ensuing discussion, it is the limitations of many of the techniques that will be emphasised, but this does not mean that they are of little value. Every technique can contribute something to generalisations made about the microflora.

Detection of Form and Arrangement of Micro-organisms in Soil

Observation of micro-organisms in the soil provides a basis for ecological studies since it gives direct evidence of the occurrence of microbes in particular environments. Earlier (p. 10) we pointed out that micro-organisms are not spread uniformly throughout the soil and that their precise arrangement varies spatially and temporally. To show this, soil must be disturbed as little as possible during its preparation for examination. It is possible to preserve soil structure quite well by impregnating soil with gelatin[181] or resins[63]. Thin sections may then be prepared from the soil when the impregnating substance has set. Examination of the sections reveals the growth patterns of fungi and their relationship to recognisable types of organic and mineral substrates, e.g. leaves, humus, mite faeces, etc. (Plate IV). The method is not readily applicable to bacteria since they are barely visible and difficult to stain in sectioned material. Attempts to stain them before impregnation of the soil may displace them from their original sites. Furthermore, bacteria are so widely dispersed in the soil that the chances of thin sections including them are remote. An exception to this is in the root region of plants where ultra-thin sections examined by electron microscopy have shown the relationship of the bacteria to the root surface[233] (Plate IV (d)). Also Jones and Griffiths[241] were able to discern the distribution of bacterial colonies, yeast cells and spore chains of actinomycetes, in 20-30 μm thick sections of soil aggregates, after staining the soil with phenolic soluble blue.

It is easier to study the spatial distribution of bacteria and actinomycetes by looking at individual soil particles mounted on slides[501, 156]. Particles stained with phenol aniline blue are shown in Plate V (a). Unfortunately, organisms that colonise opaque particles, e.g. humus, are invisible if examined with a normal transmitted light microscope. They must be stained with a fluorescent dye, e.g. acridine orange or fluoroscein isothiocyanate, and observed with incident ultraviolet light. Many fluorescent dyes will stain the micro-organisms without killing them, so that it is possible to culture cells removed from the soil with micro-manipulators. This allows the form of the organism in the soil to be compared with its form in culture[69]. Little use has been made of this technique for it is difficult and tedious.

The light microscope is limited in its application since it has a poor depth of focus and poor resolving powers, making it difficult to see the details of soil particle surfaces and microbial structure. These limitations are not shared by the scanning electron microscope[154]. In this instrument, a continuously scanning beam of electrons is focused on the surface of soil particles, causing electrons to be emitted which may be collected and made to form an image on a screen. It is possible to resolve objects only 15 nm apart whilst retaining a good depth of focus. These advantages are well illustrated in comparable photographs (Plates V (a) and V (b)) of a bacterial colony in soil, taken with a light microscope and a scanning electron microscope respectively. Scanning electron microscopes can be used to get a clearer picture of the sites of microbial activity within the soil and to study the association of different organisms at these sites. Some algae and fungi have characteristic surface structures which may permit their identification in special circumstances, but more results are needed before the full potential of this instrument can be utilised.

So far we have emphasised the role of observational techniques in establishing the spatial distribution of micro-organisms. They may also be used to determine the changes which take place in soil over a period of time. The technique most frequently applied in such studies is the contact slide technique developed by Rossi and Riccardo[405] and Cholodney[84]. The development of micro-organisms is followed by the observation of the colonisation of microscope slides buried in soil for varying periods of time. These slides may be untreated or coated with a variety of organic and inorganic materials. Alternatively, transparent strips of organic materials, e.g. cellulose, chitin and cutin[513, 157, 158] may be buried in soil. Methods of this type have been criticised because the slides present a continuous solid surface of an unnatural material which may cause condensation of water and enhanced microbial growth. This has never been conclusively proved and it is generally accepted that the technique has provided much useful data, even though it may be difficult to extrapolate from the colonisation of glass to the colonisation of natural soil particles.

The contact slide technique is best suited to studies on organisms that are adsorbed to surfaces. For examining the bacteria that grow in the soil solution, other techniques are used. Optically flat, glass-walled capillaries, often coated internally with a film of fulvic or humic acids, can be buried in soil in an effort to simulate the conditions existing in natural soil pores. If water is present in the capillaries, rapid colonisation takes place and the growth of both attached and free-swimming forms can be closely observed[360]. Growth inside the capillary is not disturbed when it is removed from the soil so that the same capillary can be observed and reburied several times.

Preservation of the soil structure and the arrangement of organisms is not so important when the morphology of individual organisms is being investigated. Indeed, it is advantageous to separate the organisms from the soil particles. This can be achieved by grinding soil in water (Jones and Mollison[242], see p. 72), or treating soil with surface active agents, e.g. sodium pyrophosphate. The dispersed cells may be examined with the light or electron microscopes. Unfortunately, most micro-organisms possess so few morphological characteristics that it is impossible to identify them with the light microscope and rarely with the electron microscope[332]; they can only be placed in broad groups. More specific identification is only possible if soil or slides that have been buried in the soil are stained with fluorescent antibody dyes. Antibodies are prepared by injecting known micro-organisms into experimental animals and harvesting the serum containing the antibodies. After purification, the antibodies are coupled with a fluorescent dye, so that when cells in the soil are stained with the resulting conjugate and viewed under ultraviolet light, they fluoresce brightly (Plate V (c)). Cells of organisms other than those to which the antibodies had been prepared do not fluoresce or show only weak fluorescence. This is the only technique that allows the simultaneous localisation and identification of micro-organisms in the soil. So far, fluorescent antisera that react specifically with species and strains of *Aspergillus*, *Arthrobotrys*, *Bacillus* and *Rhizobium* have been prepared[433, 127, 209, 434]. Other ways of obtaining similar information about these organisms exist, but they rely on indirect isolation procedures, followed by lengthy identification programmes in the laboratory.

Isolation of Micro-organisms from Soil

Micro-organisms usually have to be isolated from the soil and grown in culture before they can be identified and their reaction to environmental factors assessed. The problems of isolation can be considered under three headings, firstly those associated with the growth medium, secondly those involved in transferring organisms from the soil to the growth medium, and finally those concerned with isolation of propagules in particular growth phases from known micro-habitats.

(*a*) *Isolation Media*

The variety of media used to isolate micro-organisms from soil is enormous. They are of two types, (i) non-selective media and (ii) selective media.

Non-selective media are probably better called broad-spectrum media since they are designed to isolate broad groups of micro-organisms rather than all micro-organisms. To isolate all the micro-organisms from soil on one medium is impossible because of the tremendous diversity of their nutrient requirements. In practice, media are prepared so that they favour the development of either fungi, bacteria, actinomycetes, algae or other autotrophic forms. Normally, fungi develop best on media with a high carbon:nitrogen ratio, e.g. Czapek Dox agar (sucrose 50 g, sodium nitrate 2·0 g plus mineral salts), whilst bacteria grow better in media with a low carbon-nitrogen ratio, e.g. nutrient agar (peptone, 5·0 g, beef extract, 3·0 g, no sugars). Actinomycetes, when compared with bacteria, are frequently able to attack more resistant and complex nutrients and are often isolated by incorporating such carbon and nitrogen sources in media, e.g. starch, casein, chitin, humic acid, etc. On the other hand, most algae are able to synthesise their own nutrients and are grown on extremely simple inorganic salt solutions[380]. However, Round[410] has pointed out that liquid media are selective against many terrestrial algae since they need a solid surface on which to grow.

Many isolation media include water extracts of soil. Extracts for the isolation of algae are prepared by dissolving nutrients from baked soil[380], whilst those for the isolation of fungi and bacteria are made by autoclaving soil and water together[227] or grinding soil with cold water. The selectivity of these or any other media can be improved by the addition of inhibitory substances. Acidification or the addition of antibacterial antibiotics, e.g. aureomycin, encourages the growth of fungi, whilst addition of antifungal substances such as actidione, nystatin, and sodium propionate allows bacteria to develop better. The careful combination of antibacterial and antifungal antibiotics may, in special circumstances, favour the development of actinomycetes[564].

Narrow spectrum or selective media are designed to encourage the growth of one or a few organisms at the expense of all the others. Media may be made selective in the following ways:

(*i*) By adding a substance used by a particular organism but not by others, e.g. cellulose, as a sole carbon source selects cellulolytic organisms[124], while *Nocardia asteroides* can be isolated from soil on paraffin-coated rods[295];

(*ii*) By omitting substances required by most organisms but not the one being isolated, e.g. leaving out all organic matter, whilst supplying ammonium ions as a source of nitrogen and energy, encourages the growth of nitrifying bacteria;

(*iii*) By altering the reaction (pH) of the medium, e.g. acidification of media with acetic acid allows the slow growth of lactobacilli whilst inhibiting many other bacteria;

(*iv*) By adding a selectively microbiocidal substance, e.g. p-chloronitro-benzene kills most fungi but allows *Fusarium* to grow;

(*v*) By altering the conditions of incubation, e.g. exposure to high temperatures allows the growth of thermophilic organisms, whilst the exclusion of oxygen favours the development of anaerobes.

These different methods may be successfully combined to isolate a very specific fraction of the microflora. It would be perfectly possible to devise a medium for the isolation of a thermophilic, anaerobic, cellulose decomposing bacterium, capable of utilising inorganic nitrogen.

Selective media are used mostly for the isolation of bacteria, rather than fungi or algae. Bacteria are difficult to identify and are often most conveniently studied by grouping them according to their biochemical properties. On the other hand, fungi and algae are easier to identify because of their morphological diversity, whilst biochemically they are more uniform. Thus fungi are all aerobic heterotrophs, whilst bacteria may be aerobic or anaerobic, heterotrophic or autotrophic. Because of this, the application of selective procedures for the isolation of specific fungi and algae is generally less profitable and less successful.

(*b*) *Transfer of Organisms to the Growth Medium*

The nature of the growth medium is only part of the problem of isolation: it is necessary to transfer organisms from the soil to the medium. This can be achieved by direct and indirect methods.

Direct isolation, using micromanipulation techniques, is difficult since micro-organisms are so small. Nevertheless, practicable methods of carrying this out have been described by Warcup[540] for fungi and Casida[69] for bacteria. It is sometimes claimed that these methods isolate a more representative selection of the microflora since one could remove examples of all the morphologically different organisms that are visible. However, isolations may be biased in favour of clearly visible organisms, e.g. dark pigmented fungal forms, whilst transparent, hyaline fungi are under-represented (Table 1). A further difficulty that is encountered with this method is that organisms transferred to growth media often fail to grow, either because they are dead or because the medium is unsuitable. Casida[69] attempted to overcome some of these problems in his method by staining the soil with the vital stain acridine orange. This made the bacteria visible and, in theory, enabled a distinction to be made between living and dead cells. Living cells only take up small quantities of the dye and appear green, whilst dead cells, lacking a functional cytoplasmic membrane, absorb large quantities of dye and stain red. Unfortunately, the colour of the cells is also dependent on the concentration of the dye solution, the nature of the cell wall and the constituents of the cytoplasm, so that clear proof of viability is not obtained.

Indirect transfer methods usually involve the preparation of a soil suspension in water or a mineral solution and the addition of this suspension or a dilution of it to the isolation medium. Incubation of the inoculated medium leads to the growth of separate microbial colonies. Often, the suspension is mixed with an agar medium, poured into a petri dish and allowed to set. Alternatively, a small volume of the suspension is spread over the surface of a plate of solid agar medium so that colonies develop only on the surface. This makes their isolation easier, but it must be remembered that the speed and extent of growth, particularly of fungi, is greater on the surface and so plates must not be kept so long before cultures are removed. If low numbers of organisms are present in the soil, as is often the case with soil fungi, it may not be necessary to prepare a suspension. Lumps of soil or pieces of plant root may then be placed on the surface of the agar or dispersed in small amounts in molten agar[539]. Waksman[534] suggested that the speed with which the fungal colonies appeared on the surface of isolation plates indicated whether the fungi had been present in the soil as spores or pieces of mycelium, but this is now thought to be unreliable because of the vastly differing growth rates of different species.

If an organism is present only in extremely small amounts in the soil, it may be necessary to pre-treat the soil before attempting isolation, for if this is not done the required cells will be overgrown by their more numerous competitors. Pre-treatment aims to increase the numbers of the required organism whilst not encouraging, or actually inhibiting, the growth of all others. This process is termed 'enrichment' and it is usually used in conjunction with a suitable narrow-spectrum medium. One object of enrichment is to reproduce, on a large scale in the laboratory, the conditions existing in a naturally occurring ecological niche. Theoretically, the closer the simulated niche resembles the real niche, the more efficient will be the enrichment. Hungate[220] has pointed out that the exact duplication of an ecological niche in the laboratory is impossible because it would require the re-duplication of all the other niches affecting the one in question. For this reason, all enrichment cultures are imperfect. As an example, we may cite the enrichment of cellulolytic organisms. Evidence has been obtained that the isolation of cellulolytic organisms is affected by the substances associated with, and the nature of, the cellulose. Thus fungi may predominate on cellophane strips whilst bacteria are usually isolated from pieces of buried filter paper[231]. Burial of more natural forms of cellulose, which is contaminated with a variety of hemicelluloses, waxes and lignin, causes the development of more than just cellulolytic forms.

Enrichment can be achieved in one of two ways. Soil can either be amended at the beginning of the experiment only, by the addition of a fixed quantity of the enriching substance, or it can be continuously amended by circulation of a nutrient solution throughout the experiment.

The former 'static' method is useful if isolation is the sole aim. A variety of soluble and insoluble materials can be added to the soil including amino acids, proteins, inorganic nitrogen, starch, cellulose and chitin. Alternatively, soil may be added to water and the resulting mixture baited with such things as hemp seeds and insect wings which cause the local enrichment of aquatic phycomycete fungi. Another

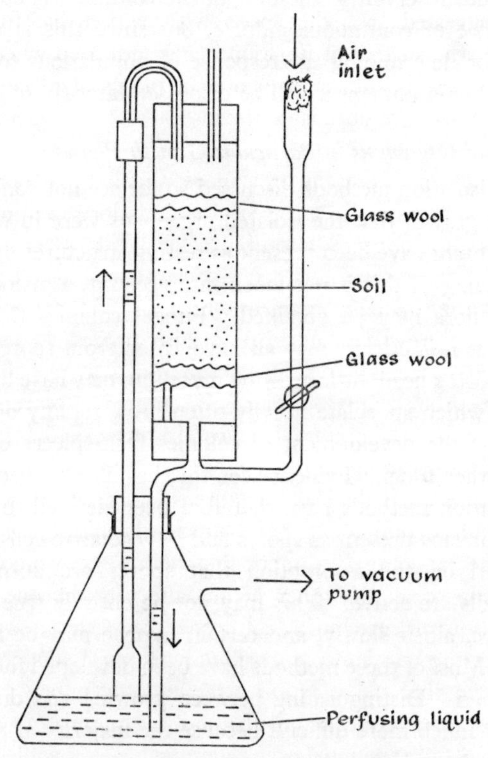

Fig. 7. *Diagram of a percolation apparatus in which a nutrient solution is continuously circulated through a column of soil. (Lees and Quastel*[263]) (*Reproduced in modified form by permission of H. Lees and* The Biochemical Journal)

bait is hair, which may be added to either soil or water for the isolation of keratinophilic organisms[166].

'Perfusion' or 'percolation' techniques, in which nutrients are continuously circulated through the soil, can also be used for isolation. In addition, they are suitable for measuring the rate of development of a population or process in soil since samples can be removed from time to time without drastically altering the system[263]. Thus perfusion techniques have been used for measuring the rate of nitrification in soil as well as the rate of production of antibiotics in amended soils. A typical perfusion system is shown in Fig. 7. An aerated nutrient solution is passed through a soil column and the samples are removed from the liquid reservoir at the base of the column. A closely related method is that of continuous culture, but since this is used almost exclusively for determining the response of populations to changes in nutrients and environment it will be discussed later (p. 69).

(c) Isolation of Organisms in Particular Growth Phases

Most of the isolation methods discussed so far are not concerned with the phase of growth that the isolated organisms were in whilst in the soil. They might have been present as resting structures or as growing vegetative cells. If this is not taken into account, a distorted picture of the microflora may be obtained. Fungal colonies that dominate dilution plates numerically may all have arisen from spores developed in a single sporing head; little growing mycelium may have been present. Since fungi which sporulate heavily often grow rapidly on agar, they may suppress the development of colonies from pieces of mycelium formed by other fungi. Evidence for this has already been discussed (p. 3). Isolation methods have therefore been devised to distinguish between organisms present as spores and as vegetative cells. They are usually based on the assumption that spores are dormant whilst vegetative cells are active. This may not be entirely true since some spores respire, albeit slowly, and certain hyphae may be inactive but still viable. Most of these methods have been developed for the examination of fungi. Distinguishing between growing and dormant cells of bacteria is much more difficult because the majority of them do not form any special resting cells.

The methods that have been developed fall into three groups; killing methods, growth methods and washing methods.

(i) Killing Methods Drying the soil and comparing the fungal colonies that developed on isolation plates before and after drying has been used to determine which species were active. McLennan[303] argued that the differences between the populations which developed were due to the death of hyphae, which were less resistant to desiccation than the spores. However, Warcup[542] showed that drying killed some fungal spores and so the accuracy of the method is suspect. Probably, desiccation of samples is only useful to suppress unwanted,

sensitive organisms, e.g. bacteria, when attempting the isolation of comparatively drought-resistant forms, such as actinomycetes present as spores.

(*ii*) *Growth Methods* In these methods, substrates are introduced into the soil but are separated from the soil particles by an air gap. It

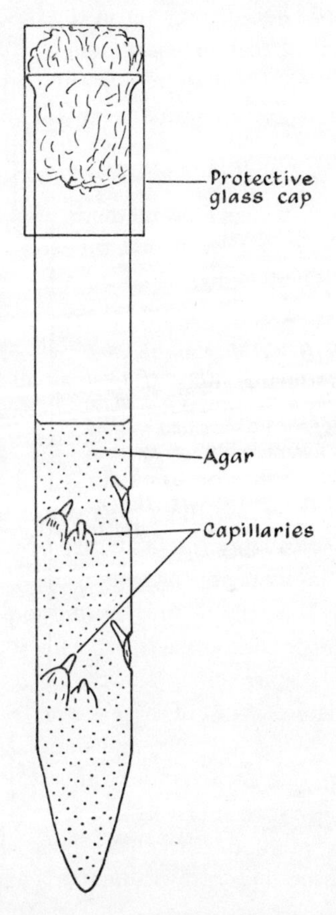

Fig. 8. *An immersion tube ready for burial in soil (Chesters*[81]*). After burial, fungi present in the soil as hyphae will grow through the capillaries and into the agar, from which they can be isolated (Reproduced by permission of C. G. C. Chesters and the British Mycological Society)*

is assumed that the fungi which colonise the substrate have grown across the gap and are therefore vegetative. The substrates are usually suspended in agar inside special containers, e.g. immersion tubes[81] (Fig. 8) or screened immersion plates[608]. Unfortunately, the introduction of these containers into soil alters the aeration and moisture characteristics and may induce germination of spores. Soil animals also carry fungal spores on their bodies and may introduce them into the agar. Nevertheless, a wider range of fungi is obtained when these methods are used together with conventional plating techniques.

(*iii*) *Washing Methods* If soil particles or roots are agitated vigorously in water, saline or dilute solutions of surface active agents, spores which are readily detachable are removed while the more

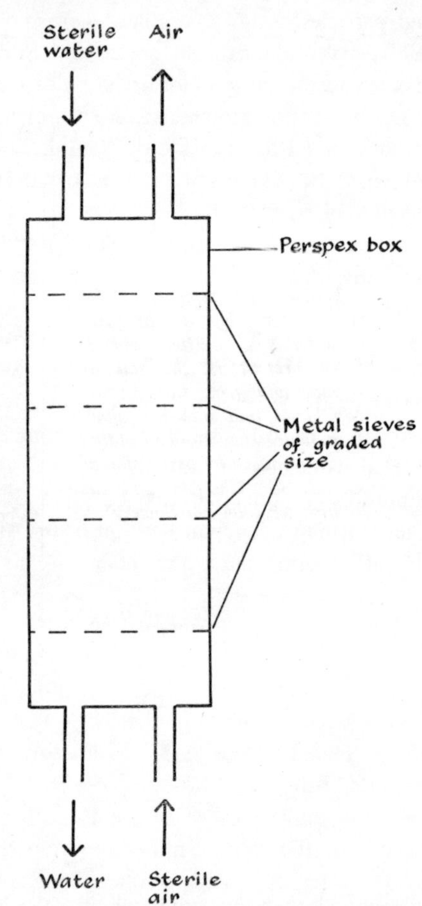

Fig. 9. *A diagram showing the operating principles of a soil washing box. Soil is placed on the uppermost sieve and washed by countercurrents of air and water. Soil, graded into different size groups, may then be plated, together with spores in the effluent water*

firmly attached fungal mycelium remains. This increases the chances of isolating fungi present as mycelium, slower growing forms, and forms inside organic particles; when the washed particles are plated out, the fungi which develop are assumed to have been active. Several washing procedures have been used. Harley and Waid[187] washed roots in many changes of water in small vials while Chesters[80] and later Williams, Parkinson and Burges[566] used special washing boxes containing sieves of varying mesh size (Fig. 9). The boxes containing the soil are filled with water and air is bubbled through, so that the soil particles are washed and, at the same time, sorted into different size groups. By plating out the washing water as well as the different sized particles, a more complete picture of the microfungi in soil and their distribution can be obtained. Surface sterilising agents, e.g. ethanol and mercuric chloride, are also used to wash roots, seeds etc. All the organisms growing from specimens treated in this way should have originated from inside the sample, implying that they are associated with the active colonisation of the substrate. Sometimes, the efficiency of surface sterilising techniques can be improved by subsequent maceration of samples. This increases the chance of isolating organisms in pure culture when a high concentration of different species is present within the material.

Detection of Microbial Activity in Soil

The methods described so far do not attempt to show how active an organism is in soil. This can be estimated by determining the rate of mycelial extension or cell division in soil, by measuring the rate of respiration (as judged by oxygen uptake and carbon dioxide output), by measuring the enzyme content of soil, or by determining the rate of substrate disappearance and metabolite accumulation.

(*a*) *Mycelial Extension and Cell Division*

The rate of mycelial extension through soil is not easy to measure unless the fungus to be studied is inoculated into sterile soil, or soil previously free of that organism. Soil may be placed in a growth tube and inoculated at one end. From time to time the soil is sampled along the tube and the rate of spread determined. Alternatively, sterile soil can be placed in a petri dish, inoculated centrally and the rate of outgrowth determined by transferring soil to a medium in another petri dish with a multipoint inoculator[486]. Plate VI (a) shows how a culture of *Streptomyces* spread during 7 days. A colony has formed at each point that the growth had reached in the soil.

Unfortunately, sterilisation of soil is difficult to achieve without altering the soil chemically and physically. Firstly, the dead organisms themselves will contribute new chemicals[12] and secondly, there will be a change in the other constituents, e.g. an increase in extractable nitrogen, phosphorus and sulphur[126]. Steam sterilisation causes the most drastic changes, but even the best methods, such as gamma radiation and methyl bromide sterilisation, cause some changes. Methyl bromide sterilisation is the least drastic since it causes less disruption of the dead organic matter. These changes will affect the rate of spread of micro-organisms, making it probable that the rates observed are far higher than those found in natural, unsterilised soil.

In addition to rate of spread, it is also necessary to determine rate of cell production. These two processes are not necessarily related since one can get good spread with or without dense growth. A possible method for measuring cell production has been suggested by Brock[51] for marine bacteria. He dissolved radioactive thymidine in sea water in which the morphologically recognisable bacterium *Leucothrix mucor* was growing. The uptake of the thymidine label into the DNA of new cells was estimated autoradiographically by counting the number of labelled cells after varying periods of incubation; from this data the rate of cell growth could be calculated.

Other substances which might be useful as labels for growing cells are the fluorescent brighteners. These vital stains are transmitted from parent to daughter cells, remaining visible for about twenty generations[102] and can be used to determine the rates of production of marked cells of bacteria, fungi and actinomycetes (Plate VI (b)).

(*b*) *Oxygen Uptake and Carbon Dioxide Output*

The overall metabolic activity of the microflora in soil may be indicated by estimates of the oxygen uptake and carbon dioxide output of soil samples. To review all the methods for making such measurements, together with their associated problems, is beyond the scope of this book. A brief outline will be given and readers are referred to Stotzky's[487] more complete review for further details.

When organic matter is attacked by micro-organisms, the following reaction takes place:

$$(CH_2O)_x + O_2 \longrightarrow CO_2 + H_2O + \text{intermediates} + \text{cellular material} + \text{energy}$$

Only 60-80 per cent of the carbon is converted to carbon dioxide, even under fully aerobic conditions because of incomplete oxidation of the substrate, which gives rise to intermediates, and the synthesis of cellular materials. Fungi and actinomycetes use a greater percentage of metabolised carbon for growth than do bacteria[6]. Theoretically, one molecule of carbon dioxide is liberated for every molecule of oxygen taken up, i.e. the respiratory quotient (RQ) is one, but this rarely happens in soil for several reasons. For example, the carbon: oxygen ratio of the substrate can alter the RQ, whilst in anaerobic respiration, no oxygen is taken up. Furthermore, carbon dioxide may be liberated chemically from the soil by the action of microbe produced acids on soil carbonates and conversely, carbon dioxide and oxygen may be bound in the soil water and not liberated. Despite these difficulties, measurement of respiration is the most frequently used method for assessing the activity of microbial populations. An example of its use is given by Parkinson and Coups[350] who showed that respiration rates could be correlated with organic matter content, moisture content and the nature of the micro-organisms in soil.

Measurements of respiration can be made in the field or the laboratory. Field measurements are made by analysing given volumes of the soil atmosphere, usually obtained by pumping air from the soil over a period of time, or by determining the amount of carbon dioxide evolved into an enclosed space above the soil surface. Witkamp[570] inserted plastic bottomless boxes into the soil to a depth of 2·5 cm. These were left in position permanently, but when measurements were taken, tight-fitting lids were attached and carbon dioxide absorbed in caustic potash in a petri dish placed inside the box. There are several problems associated with these methods, including the impossibility of distinguishing microbial respiration from root respiration and of determining the volume of soil from which the carbon dioxide came. Furthermore, in those methods involving the pumping of air from the soil, it is not known whether the induced air flow through the system affects respiration rates.

Laboratory measurements of respiration are made by removing known volumes of soil and placing them in containers which are incubated under controlled environmental conditions. To obtain reproducible results, it is necessary to sieve the soil, air dry it and then remoisten it to a known moisture level. This gross disturbance undoubtedly affects the rate of respiration for unusually high rates of respiration are often observed at the beginning of such experiments, and as we saw in Chapter 2, Greenwood[163] has suggested that this is due to spatial and chemical alterations in the soil organic matter induced by disturbance. After a period of equilibration, the rate of respiration drops to a constant but low level. Often, the amounts of carbon dioxide liberated are so small that it is difficult to distinguish between respiration rates in samples of different soils. Consequently, soils are often amended with readily available organic matter, e.g.

glucose, in order to assess the 'potential' respiratory activity in soil, rather than the actual activity.

In all of these experiments, it is not possible to distinguish between the respiration of the different micro-organisms and the soil animals. Attempts to selectively suppress the latter have been unsuccessful since many inhibitors are inactivated in the soil and some are actually metabolised. In addition, removal of one group of organisms gives other groups a greater share of the available substrates. The effects of these difficulties on attempts to measure energy flow through different parts of the soil population are discussed in Chapter 10.

Carbon dioxide is usually estimated by absorbing it in alkali and titrating the resulting solution against acid, whilst oxygen uptake is measured by following changes in manometric pressure in a closed system. The Warburg flask is a suitable container for soils in which to carry out these determinations. However, the oxygen present inside the normal Warburg flask is quickly utilised and carbon dioxide accumulates, both changes leading to alterations in the respiration rate. Use of larger flasks partly overcomes this problem, but a more effective although complex way of carrying out long term experiments is that devised by Swaby and Passey[300]. They incubated soil inside a large container in which the oxygen concentration was maintained by periodic addition of oxygen from the electrolysis of water.

(c) *Enzyme Content of Soil*

During respiration, the oxidation of organic substrates is coupled with the reduction of molecular oxygen. This is achieved by a series of enzymatic reactions termed the electron transport system (Fig. 10). One of the primary events in this system is the removal of hydrogen from the substrate by the action of dehydrogenase enzymes. Activity of micro-organisms can be assessed therefore by measuring dehydrogenase activity in soil[264] after it has been extracted from the organisms. Dehydrogenase activity is measured by reacting the enzymes with tetrazolium dyes which are converted to insoluble red coloured formozan compounds; the intensity of the red colour can be related to enzyme activity. Stevenson[479] showed that there was a significant correlation between oxygen uptake and dehydrogenase activity, but no correlation with bacterial numbers. This is not surprising since bacteria are not the only organisms producing dehydrogenase. However, Casida, Klein and Santoro[71] did show that amendment of soil with nutrients increased both bacterial numbers and dehydrogenase activity. Probably the biggest difficulty in using this technique is obtaining efficient extraction of the enzyme from the micro-organisms and then from the soil.

Activity of specific fractions of the soil microflora has also been

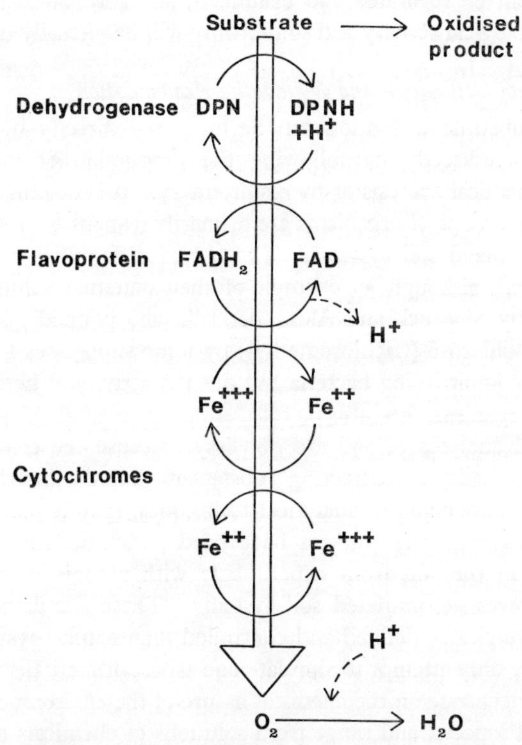

Fig. 10. *The electron transport system (Stanier, Doudoroff and Adelberg*[467]*).* DPN = *diphosphopyridine nucleotide, FAD = flavin adenine dinucleotide*

(Reproduced by permission of R. Y. Stanier and Prentice Hall Inc.)

measured by determining the activity of a variety of other enzymes. Skujins[453] has reviewed the techniques available for assay of many enzymes, including oxido-reductases, transferases, hydrolases and lyases. He has concluded that very rarely can enzyme activity be correlated with soil fertility and microbial activity because enzyme activity in soils, as measured in the laboratory, is a manifestation of several biological parameters in soil. These include free enzymes adsorbed on soil particles, free enzymes released from lysed micro-organisms, enzymes accessible in dead but not lysed cells, free enzymes released from plant roots and soil animals and metabolic activities of living roots, microbial cells and soil animals. Until these different activities can be separated and examined, any correlation of enzyme activity, biological activity and soil fertility will be virtually impossible.

(d) *Substrate Utilization and Metabolite Accumulation*

Rates of substrate utilisation can be measured directly by chemical analysis or indirectly by following the production of metabolites, again by chemical analysis or by respirometry. It is not easy to determine which groups of organisms are primarily responsible for chemical changes in mixed soil systems. The use of differential inhibitors is again limited, although an example of their potential value has been suggested by Shattuck and Alexander[443], who pointed out that the inhibitor 2-chloro-6-(trichloromethyl) pyridine suppresses the activity of autotrophic nitrifying bacteria but not the activity of heterotrophic nitrifying organisms (see also p. 160).

Chemical analyses of soil are not easy, because the errors arising from the difficulty of extracting substances, and the interference of many soil components in analytical techniques, may be considerable. For this reason, many workers have tried to predict the activity of organisms in the soil from experiments with organisms in pure or mixed cultures in simulated soil systems. These simulated systems can be more clearly defined and controlled than natural systems since they usually only attempt to simulate one aspect of natural soil. Some systems concentrate on the chemical nature of the environment, others on physical aspects, and range from solutions of chemicals thought to be in the soil to dispersion of chemicals in a solid-liquid-gas system made with grits or glass beads.

Pochon[368] has described a method for estimating the activity of soil micro-organisms in a chemically simulated culture. A series of soil dilutions are inoculated into media containing the chemical to be studied. From time to time, samples are removed and analysed for the presence or absence of the substrate and its breakdown products. The activity of the soil population in bringing about these changes is expressed in graphical form by plotting time against the first dilution in which the substrate has disappeared, or in which the metabolites have appeared (Fig. 11). Artificially high concentrations of chemicals must be used in this type of experiment (batch culture), otherwise it is impossible to detect changes. Furthermore, as the experiment progresses, the numbers of organisms present increase while the substrate concentration decreases, and so the conditions of the experiment are not constant.

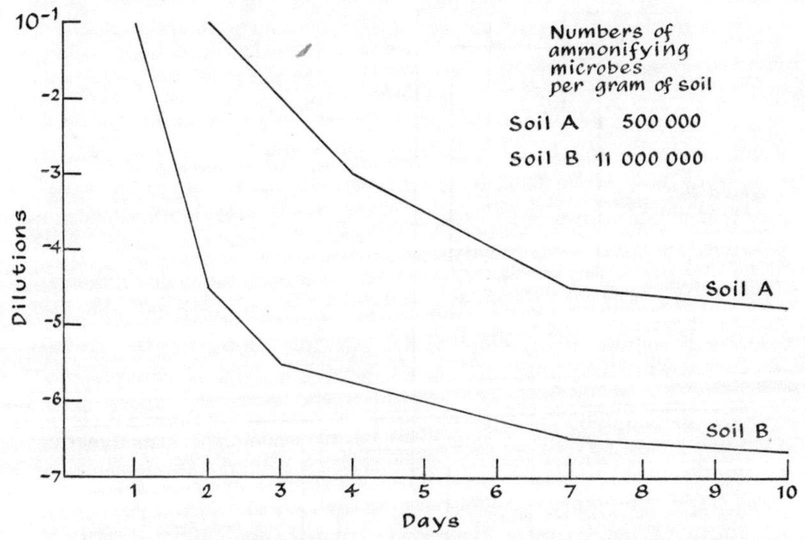

Fig. 11. *The time taken for ammonia to appear in tyrosine solutions inoculated with different dilutions of two soil suspensions. The results are plotted to show the potential activity of the soil microflora in ammonification. Both soils were cropped with flax, although the soil B was more fertile than soil A (Pochon*[368])

(*Reproduced by permission of J. Pochon and the Société Belge de Pédologie*)

One method of overcoming these problems is to carry out experiments with micro-organisms grown in a chemostat (Fig. 12). In this apparatus, micro-organisms are grown in a container to which nutrients are continuously supplied at a low flow rate but with one at a growth-limiting concentration. If the culture volume and inflowing substrate concentration are kept constant, the growth rate of the micro-organisms can be varied by varying the flow rate. If the flow rate decreases below a certain level, the cells will stop growing and may even die. If the flow rate is too high, cells will be washed out of the apparatus. It is possible, therefore, to determine the growth rates and physiological characteristics of organisms grown in a range of ecologically realistic concentrations of nutrients. Continuous culture apparatus of this type can be used as a model for some aspects of the soil environment, especially for simulating the effects of nutrient limitation, since it is somewhat similar to a soil through which drainage water is percolating. However, the model is by no means perfect since conditions in soil rarely remain so stable for such long periods of time. A more accurate

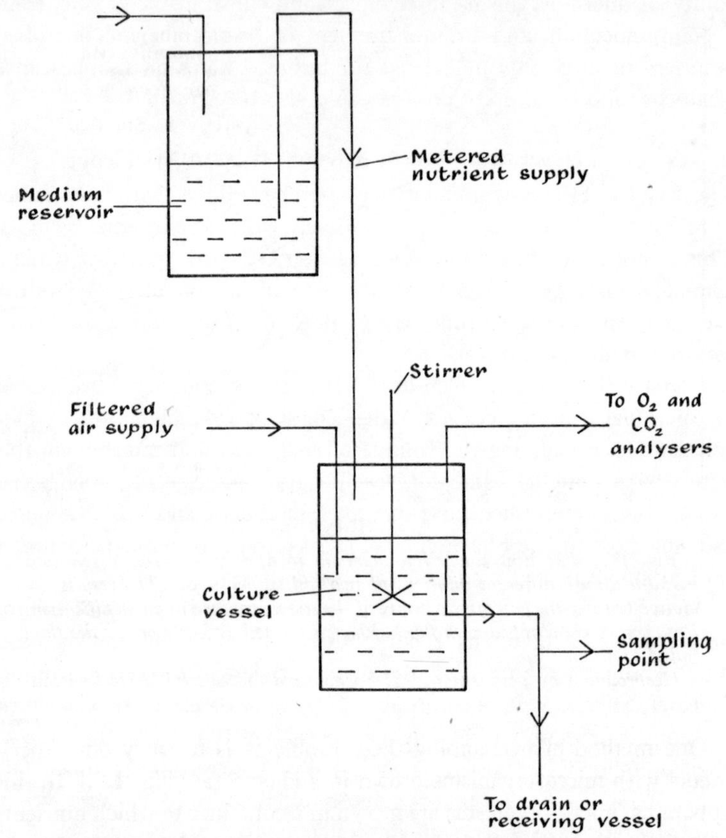

Fig. 12. *Diagrammatic representation of a chemostat in which growing micro-organisms are kept in a nutritionally constant environment* (*Rose*[401]) (*Reproduced in modified form by permission of A. H. Rose and Butterworths*)

model would be a batch system of cultivation using only low nutrient concentrations, but as we have pointed out, the changes that would take place would be small and rapid so that they would be difficult to detect. Even this model would be inadequate, because it would not take account of the solid substrates and solid matrix in which these substrates are dispersed.

Attempts to simulate this aspect of the soil environment have been made by Griffin[167, 172] who used artificial soils made of aluminium oxide grits of known particle sizes; and Parr, Parkinson and Norman[355] who used glass micro-beads. Activity in these media was assessed by ability of micro-organisms to colonise solid substrates, e.g. hair, from a point inoculum, and by measurement of respiration and microbial numbers in uniformly inoculated conditions. Rates of disappearance of glucose and build up of protein could also be estimated.

Determination of Microbial Biomass

The activity of the soil microflora has often been linked with the number of micro-organisms in a soil or occasionally with the microbial biomass. This is not justified, because not all living cells are equally active and some are dormant. Nevertheless, large microbial populations indicate the occurrence, at some time past or present, of favourable conditions for growth and/or survival.

Biomass is most conveniently determined by counting the number of microbial cells or measuring the length of mycelium present in a soil sample, calculating the volume of such cells and multiplying this figure by a notional value for specific gravity. Specific gravity has usually been determined for cells grown in culture and is often quoted as being close to 1·0, but preliminary measurements in our laboratories on actinomycete spores suggest values nearer 1·5.

Most workers have been content to determine numbers of micro-organisms in soil and have not bothered to work out the biomass. This is unfortunate for it gives a distorted view of the relative importance of different microbial groups. The ratio of numbers of bacteria to fungi in an agricultural soil may be about 100:1, but the ratio of their weights is less than 1:1. Usually biomass and numbers are expressed as a value per unit weight or volume of soil. Per unit weight estimates are used when soils of similar density are being compared, whereas per unit volume estimates are used when comparing soils of different densities, e.g. organic and mineral soils. Sometimes, biomass per unit weight of 'organic matter' or 'organic carbon' is determined. Since most of the microflora is heterotrophic, large numbers of organisms per unit weight of organic matter may indicate that a supply of nutrients and energy is readily available. Biomass may also be related to the surface area of the particles colonised; this is particularly useful in comparisons of the colonisation of root surfaces and rhizosphere soil.

Methods for counting micro-organisms can be divided into two groups, (*a*) counting by direct observation of cells and (*b*) counting by determining the number of viable cells developing in culture.

(*a*) *Counting by Direct Observation*

Conn[91] was the first to use a direct observation method for counting bacteria in soil. He made a suspension of soil and spread a small volume of this suspension over a known area of a microscope slide. After staining it with phenolic rose bengal, he was able to see and count the bacteria. More recently, fluorescein isothiocyanate has been used to stain such smears, making it easier to distinguish between bacteria and small pieces of organic matter[367, 577]. The method is subject to considerable errors since the soil smears rarely dry out evenly. Thornton and Gray[507] eliminated this problem by first mixing the soil with a suspension containing a known number of indigo particles. The suspension was placed on a microscope slide, stained with erythrosin and the number of indigo particles and bacteria counted. The numbers of bacteria per gram of soil were determined from the ratio of bacteria to indigo particles.

One problem associated with these techniques is that bacteria may be hidden by soil particles. Jones and Mollison[242] overcame this difficulty by grinding soil with water so that the soil particles were relatively small. They also improved the accuracy of the method by mixing the suspension with a warm agar solution and placing a few drops of this suspension on a haemocytometer slide under a cover slip. When the agar had set, the film was removed, stained in phenol aniline blue and observed under the microscope. Since the thickness of the agar film and the concentration of the soil in the film were known, the number of bacteria in the original soil sample could be determined. Thomas, Nicholas and Parkinson[506] described a modification of the method for measuring the length of fungal mycelium in the soil

These methods include both living and dead cells in the total counts. Attempts to distinguish between them have been made on the basis of intensity of staining and the fluorescence characteristics of cells stained with acridine orange[495]. Unfortunately, these are not reliable criteria for the reasons already discussed (p. 57).

(*b*) *Counting Cells Developing in Culture*

The number of living cells in a soil are usually estimated by culture techniques, the most popular being the dilution plate technique. A known weight of soil is suspended in a saline solution and shaken or stirred vigorously so that the micro-organisms are detached from the soil particles and the cells in the micro-colonies dispersed. Dilutions of the suspension are made, usually tenfold, and known volumes of each dilution are mixed with an agar medium in a petri dish. After incubation, cells give rise to colonies and by counting these, the number of viable cells in the soils can be calculated. The validity of the technique depends upon several questionable assumptions and it is generally agreed that plate counts result in underestimates of the soil population. On the other hand, direct counts overestimate the numbers of soil micro-organisms. Reasons for these discrepancies have been reviewed by Skinner, Jones and Mollison[451] and Jensen[235], and are summarised in Table 11.

Table 11. *Reasons for the discrepancies between the results of plate and direct counts*

Factors causing underestimation of plate counts	Factors causing overestimation of direct counts
Clumps of cells remain aggregated or attached to soil particles	Failure to distinguish cells from stained organic particles
Cells are killed in the dilution medium	Failure to distinguish bacteria from actinomycete spores
Spores fail to germinate	Failure to distinguish living from dead cells
Adsorption of cells on pipette walls	
Selectivity of the plating medium	
Selectivity of the incubation conditions	

Bacteria in particular physiological groups may be counted using the extinction dilution (most probable number) method. An extended series of tenfold dilutions is prepared and 1 ml of each dilution is inoculated into each of several tubes of medium. After incubation, the number of tubes showing growth is recorded and the most probable number of organisms accounting for the result is calculated statistically[322]. The method is less accurate than the dilution plate count, but accuracy is not the only criterion by which to judge it. Casida[70] has used the technique to isolate and count some micro-aerophilic bacteria, resembling *Streptococcus sanguis*[148] present in soil in large numbers, but not detected by the plate count method, while Eren and

Pramer[127] have used it to count the number of propagules of predacious fungi in soil.

Some modifications of these methods are worth mentioning. Firstly, bacteriophage may be counted by incorporating bacteria sensitive to their action in the medium. Zones of lysis (plaques) on dilution plates, or clearing of cells in broth culture, are regarded as proof of the presence of bacteriophage[3]. Secondly, if the number of organisms being counted is low, it may be necessary to concentrate the suspension, rather than dilute it. This can be achieved by passing large quantities of a suspension through a bacteria proof filter, and then incubating the filter on top of a nutrient medium in a petri dish as usual[106]. Finally, if it is necessary to test the response of a population to more than one medium, agar plates can be prepared and inoculated on the surface of the medium. When the colonies have grown, they are transferred to other media by pressing a velveteen pad on to the agar and then on to the new media. This replica plating technique was devised by Ledeberg and Ledeberg[262] and used for soil bacteria by James[228].

(c) Determination of Biomass without counting Procedures

A more direct method of estimating microbial biomass has recently been suggested by Jenkinson[231]. He noticed that when a soil is sterilised with chloroform and then re-inoculated with non-sterile soil, a small fraction of the organic matter is rapidly converted to carbon dioxide. This fraction, comprising 2·3-3·4 per cent of the soil carbon, he considered to have been derived from the recently killed micro-organisms and therefore represented the percentage of carbon in the biomass.

The value of this approach cannot be assessed until it has been applied to a variety of situations. However, it is one of the most interesting new methods proposed in recent years, mainly because it is based on entirely different principles to those used in counting and it may provide a relatively simple way of obtaining figures for the total biomass, rather than the biomass of individual groups of soil organisms.

4

Soil Microflora and the Decomposition of Dead Organic Matter

Some soil micro-organisms (algae and certain bacteria) are autotrophs, obtaining their energy from light or the oxidation of inorganic compounds and assimilating carbon dioxide. However, the majority of soil microbes are heterotrophic and require a supply of organic matter, which may be obtained from a variety of sources. Live plants, animals and microbes provide nutrients for the parasitic and symbiotic representatives of the soil microflora, while their exudates and dead tissues are utilised by saprophytic soil microbes. In this chapter the importance of dead organic matter as an energy source will be examined, while other sources will be dealt with in subsequent chapters.

Dead organic matter constitutes one of the most widely distributed sources of energy in the soil environment. Plant litter arriving at the soil surface is immediately attacked by soil animals and microbes so that new substrates are formed and substances released which may pass down the soil profile. At the same time, plant roots are growing through the soil, releasing soluble substances and adding gradually to the mass of particulate dead organic matter. Less is known about the utilisation of the invisible supply of soluble nutrients by soil populations than about the microbial decomposition of particulate organic matter and it is the latter which is the main subject of this chapter.

The Origins of Dead Organic Matter in Soil

(*a*) *Plant Material*

Much dead plant material is supplied to the soil in the form of leaves, twigs and bark. The total litter fall in a tropical forest may reach 153,000 kg/hectare, while in a temperate climate it may be about 25,500 kg/hectare[60]. In some cases the total litter fall may be much less and Kendrick[249] estimated that in an English pine forest it was only 4,760 kg/hectare, of which 4,080 kg was leaf litter and the remainder twig, cone and bark fragments.

Various types of plant litter were recognised by Müller[327] on the basis of their rates of incorporation into the soil. The slowly decomposing litter, forming a layer several centimetres thick over the soil surface, was termed 'mor' and is characteristic of coniferous forests. Examination of the litter (or A_0 horizon) in a pine forest shows several distinct layers, each containing needles at different stages of decomposition. The L layer at the surface consists of needles recently fallen and still intact; beneath this in the F_1 layer, needles are still recognisable but in small fragments while in the F_2 layer fragmentation has proceeded further and the needle mass is more compacted. Finally in the H (humus) layer, only amorphous organic material is present. In this system it is possible to relate the vertical position of a needle to a time scale and Kendrick and Burges[250] have estimated that a *Pinus sylvestris* needle spends about 6 months in the L layer, 2 years in the F_1 and 7 years in the F_2 before it becomes incorporated into the humus layer. However, throughout this process some soluble components are leached out of the leaves by rain and may get into the soil quite quickly.

In contrast, the bulk of 'mull' litter, which is often found in deciduous forests, is quite rapidly incorporated into the soil, although the precise rate of incorporation depends partly on the leaf type. Bocock *et al.*[35] showed that while leaves of ash (*Fraxinus excelsior*) were decomposed and incorporated into soil within 6 months, oak (*Quercus petraea*) leaves were still visible on the surface 2 years after leaf fall. It is possible for deciduous leaves to form a mor litter if the underlying soil conditions slow down decomposition. This was shown experimentally by Bocock and Gilbert[34], who found that leaves of birch (*Betula verrucosa*), lime (*Tilia cordata*) and other trees broke down more slowly on mor sites than on mull.

The formation of mull and mor litter has interested workers for many years and their results have shown that a number of interrelated factors are involved.

(*i*) Climatic Factors. Lower temperatures and high rainfall cause slower rates of decomposition and therefore favour the development of mor litter.

(*ii*) Edaphic Factors. Soil conditions, especially the base content, influence the susceptibility of leaves to decomposition, high base contents favouring mull formation.

(*iii*) Biotic Factors. Higher activity of soil microbes and animals (especially earthworms) leads to more rapid decomposition, and hence promotes mull formation.

(*iv*) Chemical Factors. Differences in the chemical composition of leaves influence their resistance to decomposition.

Some of the most interesting work on the interrelationships between these factors has been concerned with the chemical constituents of leaves. Handley[183] suggested that stabilised proteins, resistant to decomposition, occurred in leaves forming mor litter. He later obtained complexes *in vitro* between extracts of various leaves; the most resistant complexes were formed with extracts of mor-forming leaves[184]. Coulson, Davies and Lewis[97] found that beech leaves (*Fagus sylvatica*) from trees grown on a base-deficient soil (mor site) had a higher content of polyphenols than those from a base-rich mull site and Davies, Coulson and Lewis[107] suggested that the polyphenols were responsible for the stabilisation of leaf proteins. Basaraba and Starkey[24] found that plant tannins could form complexes with proteins which also made them more resistant to microbial attack. There is therefore good evidence for the interaction between soil conditions and the chemical composition of leaves determining their subsequent rate of decomposition by micro-organisms.

Another source of dead organic matter in soil is plant roots. A brief examination of a soil will show that many small detached root fragments are present, but unfortunately it is difficult to obtain measurements of their rate of supply. The root : shoot ratio of living plants suggests that, on average, about half as much root material enters the soil as compared with surface litter, but these estimates do not take into account the continuous supply of dead cells and root exudates from living roots which it is not possible to measure (see Chapter 6). It is known that seasonal changes occur in the quantities of roots in soil[66]. The length of roots of *Pinus sylvestris* doubled from May to July (500 to 1,000 m/m^2), but by October the amount had returned to 500 m/m^2, indicating a considerable supply of dead root material to the soil[243].

(*b*) *Animal and Microbial Material*

It is often assumed that the contribution of dead organic material from animals and microbes is small in comparison with that from plants. However, many of the decomposition products of plant material are incorporated into animal and microbial cells so that complete mineralisation does not occur until these organisms have, in turn, been decomposed. Some idea of the importance of these sources of organic matter can be obtained by considering their biomass in soil. The biomass of the main invertebrate groups in a temperate grassland soil was estimated to be about 2,300 kg/hectare, while Clark[86] calculated that there were 4,500 kg/hectare live weight of bacteria in an arable soil, together with the same weight of fungi.

Animals are usually mobile and may migrate into any part of the soil, although they are most concentrated in litter layers, whilst microbes are present throughout the profile. When they die, therefore, new sources of organic matter will be made available both within the litter layer and the mineral soil. In addition to these supplies that originate within soil, dead animal remains and fruit bodies of higher fungi arrive at the surface and are decomposed and incorporated into the soil like plant litter.

Succession of Micro-organisms

'Succession' is an ecological term describing the sequence of organisms occurring in a given time over a defined area. Two broad types of succession, autogenic and allogenic, are recognised in communities of higher plants. In an autogenic succession changes are brought about by each wave of colonisers, ameliorating the environment for the next wave; in other words, the activities of the plants themselves largely control the sequence of events. In an allogenic succession, factors not under the control of the plants (e.g. climate) determine the pattern of change. In most plant communities both types of succession operate, often simultaneously, and it is difficult to distinguish between them. This is even more difficult in examples of microbial succession.

Within plant communities, succession of soil microbes closely related to higher plants can sometimes be detected; this phenomenon in sand dunes was demonstrated by Webley, Eastwood and Gimmingham[554], where roots of the dune plants had a marked influence on the composition of the microbial population. In such a case, the succession of the micro-organisms is governed largely by external factors (i.e. the plant roots) and is therefore allogenic. However, within a soil, organic substrates are colonised by successive waves of micro-organisms, each wave altering the micro-habitat and making way for the next. This situation closely resembles an autogenic succession of higher plants, except that there is a progressive depletion of the chemical energy sources by the heterotrophic organisms[142]. It is this type of succession that concerns us here.

(*a*) *Succession on Plant Material*

Plant tissues do not die suddenly but have a period of senescence during which they are almost invariably attacked by weak parasites[143]. Therefore, the pioneer colonisers on plant material recently added to soil are not typical soil saprophytes, but more specialised microbes which can overcome the declining resistance of the tissues. These organisms are then displaced by the true soil microflora. Various studies of the succession of fungi on plant debris have clearly demonstrated this pattern; similar data for other soil organisms are lacking because of difficulties in their detection by direct observation and identification.

(*i*) *Leaf Litter* The colonisation of dead leaves by fungi has been studied by isolating organisms from and direct observation of leaf litter at various stages of decomposition and a summary of some of the results obtained is given in Table 12. The colonisers of freshly fallen leaves are generally rather uncommon species (often ascomycetes, basidiomycetes or pycnidia-forming species) and some of them have never

Table 12. *Colonisation of decomposing leaf litters by fungi*

Initial colonisers of freshly fallen leaves	Colonisers of material incorporated into upper layers of soil	Duration of decomposition process	Type of leaf
Phomopsis scobina *Cloeophora rhododendri* *Phoma* sp.	*Trichoderma viride* *Penicillium* spp.	12–18 months	*Fraxinus excelsior* (ash)[205]
Polyscytalum fecundissimum *Aureobasidium pullulans*	*Trichoderma viride* *Penicillium* spp.	18–24 months	*Quercus petraea* (oak)[205]
Protostegia eucalypti *Cladosporium herbarum* *Readeriella mirabilis*	*Trichoderma viride* *Penicillium* spp. *Mucor* spp. *Rhizopus nigricans* *Cunninghamella echinulata*	15 months	*Eucalyptus regnans* (Eucalyptus)[285]
Lophodermium pinastri *Fusiccocum bacillare* *Desmazierella acicola*	*Trichoderma viride* *Penicillium* spp. *Mortierella* spp.	> 7 years	*Pinus sylvestris* (pine)[250,565]

been detected on any other substrate. Many of them are probably weak parasites and indeed some (e.g. *Lophodermium pinastri* on *Pinus sylvestris*) infect the leaves while they are still on the tree. Others may be part of the surface microflora of the living leaf, constituting the 'phyllosphere' population[421], which probably initiates the decomposition of ageing leaves. Much of the initial colonisation is confined to internal tissues but once the leaf has entered the more humid sub-surface layers, external growth increases.

Over the period of decomposition, which ranges from 12 to 18 months for *Fraxinus*, to over 7 years for *Pinus*, a gradual transition of the microflora occurs, more specialised colonisers with a low saprophytic ability being replaced by typical soil saprophytes. When the material is fully incorporated into the upper layers of the soil, it is colonised by soil fungi such as *Penicillium* spp., *Trichoderma viride* and various Zygomycetes. By this stage, the microflora associated with the different types of leaf debris is almost identical, in marked contrast to the diversity of the initial colonisers.

The invasion of the leaves by the soil microflora is assisted considerably by the activities of soil animals. In addition to their well-known capacity to incorporate surface litter into soil, they also assist in the physical breakdown of the leaves, causing exposure of fresh surfaces to microbial attack and providing substrates in the form of faeces. Their importance was clearly demonstrated by Edwards and Heath[123], who placed leaf discs of oak (*Quercus*) and beech (*Fagus*) in nylon bags with various mesh sizes in soil. Over a period of 9 months, the reduction in area of the discs was measured. In all the bags into which soil animals penetrated, disappearance of leaf tissues was observed, mostly from bags permitting entry of earthworms. When the mesh excluded all animals, no visible breakdown of the leaves occurred.

(*ii*) *Roots* It is a more difficult task to follow the succession of microbes on roots as they die in soil, since it is impossible to define an exact starting-point comparable with leaf fall or to relate the spatial arrangement of root debris to a time scale for decomposition. As a result, methods used to study root decomposition differ from those applied to leaf litter.

Sometimes living roots are completely removed from soil or surface-sterilised and then buried in soil, thus giving a defined but artificial starting-point to the decomposition process. Such methods do not give information on the initial stages of succession in senescent root tissues.

An alternative is to study the colonisation of intact roots in soil, following changes in the nature and distribution of the microflora within root tissues as they become senescent. This may be carried out on the root of whole plants or on roots excised from the aerial shoots. The decomposition of broad bean roots of intact and excised plants has been followed by Dix[113], who found that, as roots become moribund, their internal tissues are invaded by some of the fungi also found on the surface of living roots (e.g. *Fusarium* spp. and *Gliocladium roseum*). Other root surface colonisers, such as *Penicillium*

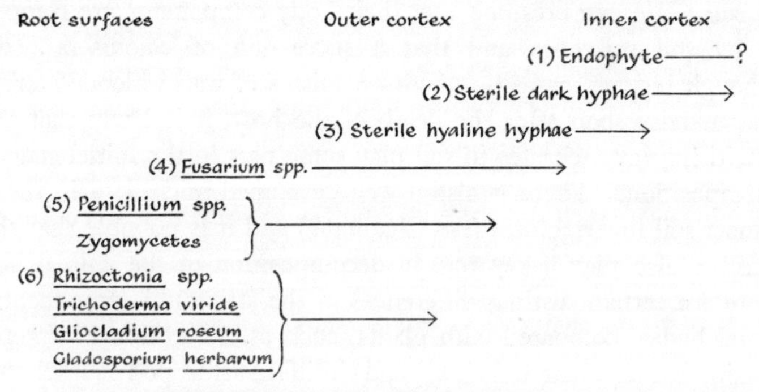

Fig. 13. *The succession of fungi in the cortex tissues of decomposing rye grass roots* (Lolium perenne)[533]

spp. and *Trichoderma viride*, decline in frequency as the roots begin to die. Excision of the roots did not produce marked changes in the nature of the fungal colonisers. A similar conclusion was reached by Dickinson and Pugh[112] when comparing the root surface fungi of intact and excised *Halimione portulacoides* plants.

Waid[533] sampled naturally occurring rye grass roots at various stages of decomposition and studied the distribution of fungi within their tissues (Fig. 13). The pioneer colonisers of the outer cortex tissues were rather specialised forms with little ability to compete with other saprophytic micro-organisms (sterile forms and an endophyte). As host resistance declined, those with greater saprophytic ability (e.g. *Penicillium* spp. and *Trichoderma viride*) replaced the earlier colonisers.

The picture of succession on roots is therefore essentially similar to that on decomposing leaves, but is complicated by the presence of a great diversity of soil microbes on the surface of 'live' roots. These range from typical soil saprophytes to specialised pathogens or symbionts (see Chapter 6). As senescence begins, internal tissues are penetrated firstly by the more specialised forms; soil saprophytes like *Penicillium*, although living on dead organic matter from living roots, do not begin to decompose internal root tissues until the resistance of the cells has declined.

(*b*) *Succession on Animal and Microbial Material*

The study of decomposing animal corpses in soil has never been a popular pastime so that there is little known about the succession of micro-organisms on animal tissues originating above ground or within the soil. We can presume that their body components are competed for by soil microbes and that a succession of colonisers occurs. Although internal parasitic micro-organisms of warm-blooded animals cease activity soon after the death of the host, it is likely that those parasitising invertebrates in soil play some part in the initial stages of decomposition. There is also a well-developed microflora in the gut of most soil invertebrates (see Chapter 9) and it is possible that these microbes also play a key role in decomposition of the animal body. There are certain distinct differences in the structural components of animal bodies compared with plants, such as the presence of keratin (in skin, feathers, hair and claws) and chitin (in arthropod exoskeletons), and it is the breakdown of such parts that has been most studied.

The succession of microbes on hair buried in soil was followed by Griffin[166]. The early colonisers were common soil fungi and he concluded that they were growing on substances associated with the keratin. Later, fungi capable of breaking down more complex substances appeared and finally the true keratinophilic fungi. These are found mainly in association with keratin and include species of *Keratinomyces*, *Trichophyton* and *Microsporum*, some of which cause skin diseases of animals and humans. They appeared on the hair between 17 and 38 days after burial, depending upon the soil type. Keratinophilic fungi have been found in a wide variety of habitats in coastal soils by Pugh and Mathison[382]. On sand dunes with a population of rabbits, hares and birds, all being potential sources of keratin, *Arthroderma curreyi* and *Ctenomyces serratus* were quite abundant. Soil from bird traps where feathers accumulated provided most *C. serratus* and the high-water mark, with an abundance of feathers and dog fish cases, gave most *A. curreyi*.

The decomposition of insect wings in soil has been investigated by Okafor[342] who emphasised the structural complexity of the insect cuticle. This consists of a thin epicuticle made up of cement, wax and cuticulin layers, and a thick pro-cuticle composed of lipo-protein and chitin (Fig. 14). The outer portion of the lipo-protein layer becomes impregnated with an amino-phenol. To elucidate the effects of the various components on decomposition, de-waxed, and de-proteinised,

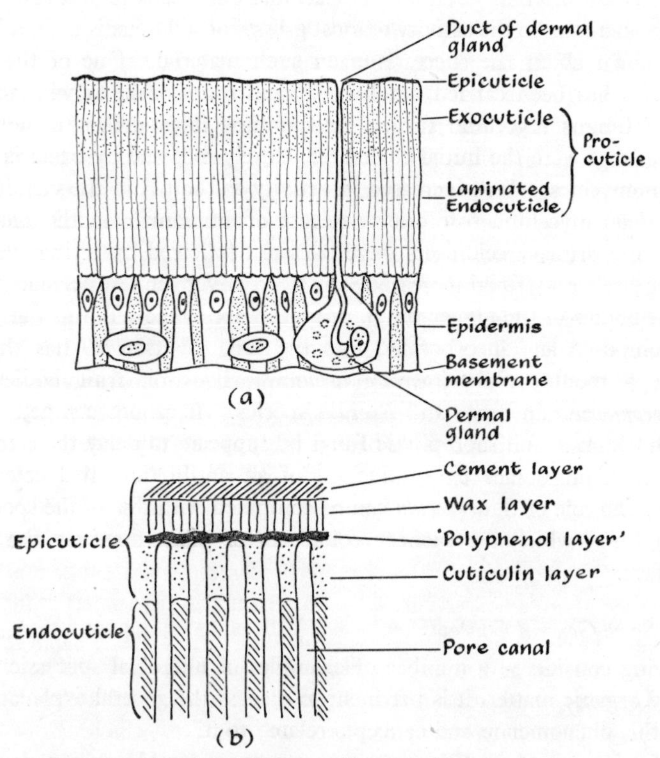

Fig. 14. (a) *Generalised section of the insect integument:* (b) *schematic section of the epicuticle* (*Wigglesworth*[559]) (*Reproduced in modified form by permission of Sir Vincent Wigglesworth and the Cambridge Philosophical Society*)

and untreated wings were buried in soil. The de-waxed, de-proteinised wings, consisting mainly of chitin, were completely decomposed and *Mortierella* sp., *Pseudomonas* sp. and two *Streptomyces* spp. were thought to be mainly responsible for the hydrolysis of the chitin. The de-waxed wings were partially decomposed but untreated ones were still intact after 300 days in soil. It was suggested that the wax layer of the epicuticle, together with the H-groups of the lipo-protein and the tanning phenols of the pro-cuticle, were responsible for the resistance of the wings to microbial attack. Thus it appears that succession of microbes on insect exoskeletons in soil is governed mainly by the differential resistance of the various layered components to decomposition.

It is known that addition of dead microbial cells to soil results in large increases in the activity of most groups of soil organisms, but little is known about the successions on such material. One of the few studies has been carried out by Mayfield (unpublished), who added dead fungus mycelium to acid and alkaline forest soil. In the acid soil, fungi were the initial colonisers, closely followed by bacteria and actinomycetes, which increased in activity as the pH in the vicinity of the dead mycelium rose due to release of ammonia. In the alkaline soil, the primary colonisers were bacteria, followed by actinomycetes after one week; fungi increased very slowly over the test period. The fruit bodies of the higher fungi, produced above the soil, must also be decomposed and incorporated into the soil. Watson[552] has shown that a mould, *Calcarisporium arbuscula*, lives on fruit bodies of *Hygrophorus*, *Lactarius* and *Russula* species. It cannot compete well with common soil saprophytic fungi but appears to delay their attack on the fruit bodies by production of an antibiotic called calcarin. Thus the role of *C. arbuscula* can be compared with that of the specialised fungi which are pioneer colonisers of plant litter on the soil surface.

(c) *Some General Concepts*

Having considered a number of examples of microbial succession on dead organic matter, it is pertinent to discuss the general explanations for this phenomenon and concepts related to it.

If micro-organism 'B' succeeds micro-organism 'A' during decomposition of a substrate, there are a number of possible explanations for this.

(*i*) Food sources available to 'A' are exhausted (e.g. sugars) and while 'B' can utilise the remaining sources (e.g. cellulose), 'A' cannot.

(*ii*) 'A' ameliorates the environment for 'B' which then becomes active. This may be brought about in a number of ways. Nutrients previously unavailable to 'B' may be altered chemically or physically to an available form, e.g. chitin is hydrolysed to amino-sugars, or the dead cells of 'A' provide a substrate for 'B'. Previously protected substrates, such as chitin in the insect exoskeleton, may be exposed by 'A' and utilised by 'B'. Another possibility is that as a result of

'A's' activity, a change in the physical environment is produced (e.g. depletion of oxygen) which decreases its own activity but stimulates 'B'. Finally, the substrate, although moribund, may still have some residual resistance to infection and while 'A' can overcome this, 'B' cannot invade the tissues until they are killed by 'A'.

(*iii*) 'A' suppresses the activity of 'B' by direct inhibition (e.g. by antibiotics), by inducing unfavourable environmental conditions (e.g. increasing carbon dioxide concentrations) or by competing successfully for available nutrient supplies (e.g. by having a faster growth rate).

(*iv*) A change in environmental conditions is brought about by factors independent of the decomposition process, e.g. heavy rain causes the build-up of anaerobic conditions around the substrate and these are tolerated by 'B' but not by 'A'. The succession in this case is clearly allogenic.

(*v*) The sequence 'A'→'B' is an artefact produced by inadequacies of the methods used to detect the activity of these microbes. Thus 'B' may develop more slowly than 'A' on the isolation medium and not be detected until 'A' has declined.

During the past 50 years, a number of attempts have been made to categorise micro-organisms on the basis of their respective roles in succession. The breakdown of any substrate reaching the soil may be regarded as a two-stage process. When fresh materials reach soil, a flare-up of organisms occurs; the more readily assimilated substances are quickly used up and the microbes return to their original state of rest or low activity. The micro-organisms participating in this first stage were termed *zymogenous* by Winogradsky[568] and include many Fungi Imperfecti and bacteria such as *Bacillus* and *Pseudomonas* species which are prominent among those using simple carbon compounds[212]. The remaining materials, consisting of both constituents of the added substrate and the products of the earlier colonisers, are more resistant to decomposition and there is then far less intense microbial activity, the situation being similar to that in soil before the addition of the fresh substrate. It is at this stage that the *autochthonus* flora[568] is important, having a low but steady level of activity and living on the native soil organic matter (or humus). Winogradsky's[568] categorisation of the soil microflora has had a considerable influence on the development of ecological theory but it is not always possible to decide in which category to place a micro-organism. Actinomycetes, for example, are often considered to be important in the later stages of decomposition and are included in the autochthonous microflora, but it is more likely that their apparent late arrival is due to methodological artefacts of the type mentioned above.

A detailed categorisation of soil fungi was made by Garrett[141], who proposed a number of ecological groups, relating biological attributes to roles in decomposition processes. Among the early colonisers of organic debris are the *sugar fungi* which include several Zygomycetes, e.g. *Absidia*, *Mucor*, *Rhizopus*, and many Fungi Imperfecti. All of these fungi are capable of rapid spore germination and mycelial growth in the presence of suitable substrates. In the presence of simple organic compounds, they flare-up quickly, utilise these compounds and then produce many spores, their mycelium dying off quite rapidly. These fungi are widely distributed in soil, their spores being particularly abundant, and are typically zymogenous. Garrett[141] also recognised the *lignin* or *humus fungi*, which include basidiomycetes capable of using complex molecules, like lignin. These have a much slower growth rate than sugar fungi and cannot compete successfully with them for simple organic compounds. Their mycelium, however, is more persistent, often forming strands or rhizomorphs (see Chapter 1) which assist in the penetration of the hard woody materials which they decompose. These fungi would be placed in the autochthonous microflora.

The ability of a micro-organism to colonise a substrate in competition with other members of the soil microflora is termed its 'competitive saprophytic ability'. This term was coined by Garrett[140] and originally applied to plant pathogenic fungi in soil, distinguishing between soil inhabitants, which have a high competitive ability, and root inhabitants, which have a lower one (see Chapter 6). Among the attributes contributing to the competitive saprophytic ability of a fungal pathogen are its growth rate, enzyme production, production of inhibitors and tolerance of inhibitors produced by its competitors. Such factors also influence the outcome of competition between completely saprophytic members of the soil microflora for any one substrate under a given set of environmental conditions. Hence the concepts developed by Garrett[140] have proved valuable in studies of both saprophytic and parasitic microbes in soil

Major Chemical Changes during Decomposition of Dead Organic Matter

Certain general trends are evident from studies of chemical changes in decomposing organic matter. The first materials to be removed, by the combined action of microbes and leaching, are water-soluble compounds, such as sodium and potassium salts and sugars. After these, more complex substances like starch and proteins are attacked. Whereas many of these components of organic matter may be decomposed within a few months, more complex structural materials, such as lignin and keratin, may persist in soil for some years. One complicating factor is the synthesis and release of substances by the organisms decomposing the organic matter which cannot always be distinguished from the components of the original material. It must also be emphasised that generalisations on the order of breakdown of components of organic matter do not always apply. As we have seen, relatively simple components may be protected from rapid decomposition by their association with other substances.

During the initial phases of decomposition of most types of organic matter, there is a marked reduction in the total carbon content. The decomposition in soil of rye grass roots and tops labelled with ^{14}C was followed by Jenkinson[231]. After 6 months, 60 per cent of the labelled carbon was gone, but from then on decomposition was much slower and about 20 per cent was still present after 4 years (Fig. 15). Initially, rye grass tops released carbon more quickly than roots, but after 1 year the differences disappeared. Chemical changes in *Abies*, *Picea* and *Pinus* litters were studied by Hayes[197]. The carbon content either decreased for the first 6 months and then showed only slight changes, or showed little change throughout the whole experimental period. The loss of carbon from the bacterium *Nitrosomonas europea* decomposing in soil was measured by introducing ^{14}C-labelled cells into soil and partially sterilising the system with chloroform vapour[231]. After 10 days, almost 30 per cent of the carbon in the bacterial cells had been liberated as carbon dioxide.

The rate of release of carbon from decomposing residues is considerably influenced by their nitrogen content. Whereas in woody tissues there may be only 0·2-0·5 per cent organic nitrogen, in softer tissues such as legume hay, 1·5-3·0 per cent may be present[21]. This results in marked differences in their carbon : nitrogen ratios. Micro-organisms decomposing the residues must have a source of nitrogen to build new protoplasm and this is obtained from the residues themselves or from other sources in soil. If adequate supplies of nitrogen are present, the carbon : nitrogen ratio of organic matter falls during the initial period of rapid carbon loss, most of the nitrogen being preserved in microbial cells associated with the residues, and in more complex recalcitrant molecules. After the ratio has reached about 20:1, the rate of change becomes much slower. If, on the other hand, a non-nitrogenous substrate such as cellulose is buried in soil, little decomposition will occur unless sufficient extraneous supplies of nitrogen are available.

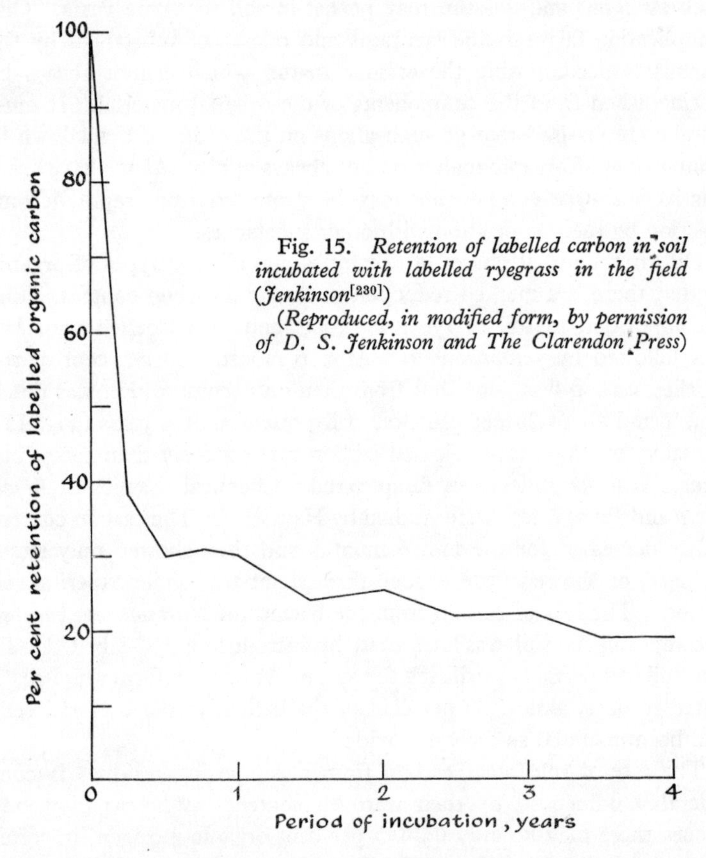

Fig. 15. *Retention of labelled carbon in soil incubated with labelled ryegrass in the field (Jenkinson*[230]*)*

(Reproduced, in modified form, by permission of D. S. Jenkinson and The Clarendon Press)

The relative amounts of nitrogen and carbon in fresh organic debris can therefore have a considerable effect on the nitrogen status of the soil. If insufficient nitrogen is present in the debris, nitrate and ammonium ions may be removed from the soil solution by microbial decomposers and hence become temporarily immobilised and unavailable for plant growth (see also Chapter 7 for other examples of immobilisation). When sufficient nitrogen is present in the substrate to support microbial decomposition, mineralisation of the substrate nitrogen will occur and plant growth will be encouraged. The balance between immobilisation and mineralisation of nitrogen also changes as the carbon : nitrogen ratio of residues falls during decomposition. This was demonstrated in experiments on the breakdown of wheat straw[14]. As the carbon : nitrogen ratio fell, immobilisation of nitrogen increased, reaching a maximum of 25 :1 after 20 days; after this, immediate release of nitrogen began and mineralisation became the dominant process.

The rate at which organic matter is added to soil also affects its subsequent breakdown, the percentage decomposition being inversely related to the rate of addition. Although the total output of carbon dioxide from decomposition increases with higher rates of addition, the percentage of added carbon evolved as carbon dioxide falls (Table 13). The reasons for this effect are not clear. Finally, it is

Table 13. *Influence of the rate of residue addition on residue decomposition*[87]

Alfalfa meal added	mg CO_2 per 50 g soil			Percentage added C evolved as CO_2		
	1 week	2 weeks	3 weeks	1 week	2 weeks	3 weeks
0	44	61	77	—	—	—
0·8%	261	360	421	35·0	48·2	55·5
1·6%	367	520	606	26·0	37·0	42·7
3·2%	532	860	1007	19·7	32·2	37·5

important to note the so-called 'priming effect' of organic matter added to soil. Addition of fresh residues to soil increases the decomposition rate of the native soil organic matter present before the amendment. Possibly the new materials provide an energy source for the autochthonous flora and enable these organisms to become more active. However, results from many laboratory experiments suggest that recalcitrant organic compounds may be attacked by micro-organisms normally regarded as zymogenous, e.g. *Penicillium* sp., providing that a readily available energy source is also present. In some cases this is due to induction of enzymes. It is therefore likely that the so-called zymogenous microflora also plays a part in the decomposition of native organic matter after additions of fresh substrate to soil[578].

5

Breakdown of Organic Chemicals in Soil

So far we have considered the role of micro-organisms in the decomposition of naturally occurring plant, animal and microbial residues. When such debris reaches the soil, many specific chemical substances become available to the colonising organisms at different periods of time. About 90 per cent of the dry weight of this debris is composed of organic materials, and some idea of the diversity and relative amounts in different organisms is given in Table 14. The greatest differences between organisms are to be found in the chemicals involved in maintaining their structure or skeleton. Plants contain a high proportion of cellulose, hemicelluloses and lignins, while in yeasts,

Table 14. *Chemical composition of some living organisms and tissues*

	Grams/100 grams dry weight			
	Yeast cells[196]	Plant tissue[6]	Mammalian muscle[557]	Insect integument[561]
Mannan and glucan	28	?	—	—
Chitin	?	—	—	25-55
Cellulose	—	15-60	—	—
Hemicellulose	—	10-30	—	—
Lignin	—	5-30	—	—
Proteins and nucleic acids	24	2-15	80	25-37
Amino acids	1·3	?	?	?
Aliphatic acids	?	5-30	4·8	?
Soluble sugars	0·5	?	?	?
Glycogen	18	—	2·0-7·2	—
Lipids	9	1-25	5·2	4-4·5
Mineral ash	17	1-13	4·4	1
Undetermined	2·2	?	?	Tanned protein and lipoprotein Pigments, phenols

mannan and glucan are important. In animals, especially those found in soil, chitin is the most important structural component and it is also to be found in the cell walls of some fungi. All living organisms contain protein and nucleic acids, the former constituting a high proportion of animal tissues, a variety of storage compounds such as lipids and the carbohydrates glycogen and starch, and a selection of soluble substances, e.g. sugars and amino acids.

The availability of these various substrates is studied by presenting the microflora with each chemical in a relatively pure form. The general impression gained from such experiments is that any organic substance can be attacked by one or more types of micro-organism under the appropriate environmental conditions. However, recent experiments on the breakdown of artificially produced organic chemicals, such as pesticides, have suggested that there may be some substances that cannot be attacked and that there are many, both natural and artificial, which are broken down very slowly even under the most favourable environmental conditions[8]. Chemicals which are comparatively resistant to attack are termed 'recalcitrant'. Molecular recalcitrance may be linked with several properties of the microflora, including the inability of organisms to produce the necessary enzymes, the impermeability of the micro-organisms to the substrate and the susceptibility of the organism's enzymes to inhibition by the substrate. It may also be due to innate characteristics of the chemical itself and it has often been observed that aromatic chemicals, e.g. polyphenols, are relatively resistant to attack and may protect other chemicals from attack if they are associated with them (see p. 77).

Much of the information in this chapter, while indicating the great potential of the microflora in causing decay, is not intended to account wholly for the disappearance of substrates in natural environments. Only a representative selection of materials have been chosen for comment, the choice being influenced chiefly by the fragmentary nature of the available data.

Decomposition of Simple (Monomeric) Organic Compounds

Many of the simpler cell constituents are very readily utilised by a wide range of soil micro-organisms. Substances such as sugars, amino sugars, organic acids and amino acids are released into the soil from cell protoplasm or are produced by the decomposition of more complex compounds. All of these materials are rapidly utilised by the soil microflora and addition of them to soil results in marked increases in microbial activity and decomposition of native organic matter.

The great majority of soil micro-organisms can utilise most organic acids produced during the oxidation of carbohydrates and proteins, so the amount of organic acids in soil is usually very small. They occur most frequently in waterlogged, partially anaerobic conditions.

Table 15. *Soluble sugars in hydrolysates of soil expressed as mg/g soil*[16]

	Soil under grass[180]	Loam[79]	B_2 horizon podol[284]
Galactose	0·8	0·1	1·8
Glucose	2·3	0·1	3·9
Mannose	0·7	0·03	1·8
Arabinose	0·6	0·2	1·0
Xylose	0·7	0·1	1·0
Ribose	0·1	0·2 (Ribose + Fucose)	0·7 (Ribose + Fucose)
Fucose	0·2		
Rhamnose	0·4	0·1	1·0
% organic matter	12·1	1·4	13·9

Takijima[503] has found acetic, formic, butyric, lactic and succinic acids in rice paddy soils while Wang, Yang and Chuang[538] have reported phenolic acids such as p-hydroxy benzoic, p-coumaric, vanillic, ferulic and syringic acids in sugar-cane fields. Stevenson and Ivarson[480] studied the decomposition of ^{14}C acetate in soil. After 6-9 hours, 22-30 per cent of added acetate had been completely oxidised to carbon dioxide. The remaining acetate could not be recovered and it was presumed that it had been incorporated into microbial protoplasm or metabolised non-oxidatively and deposited in the soil organic fraction.

Water-soluble sugars are present in the soil only in small quantities but many do occur in the form of insoluble polymers. Any organism that can attack these polymers and liberate the sugars will have a plentiful energy supply. Some idea of the range and quantities of sugars that could become available in this way is given in Table 15. Many of them are the constituent sugars of polysaccharides produced by micro-organisms[8] and hemicelluloses. Some are attacked rapidly and Chahal and Wagner[73], who studied the decomposition of ^{14}C-

labelled glucose, estimated that after 3 months, 75 per cent of it had been converted to carbon dioxide. There is evidence that the remainder is incorporated into the humus fraction of the soil[310]. Addition of sugars like glucose to soil also has a priming effect, causing an increased rate of decomposition of the native organic matter in the soil[291].

Most of the amino acids found in soil are not in a free state; they may be polymerised in the form of proteins or associated with complex humic materials. Table 16 gives some idea of the types and amounts

Table 16. *Amino acids in hydrolysates of soils expressed as mg/g soil*

	Podzol[464]	Prairie soil[464]	Loam[477]
Aspartic acid	1·9	1·2	0·5
Serine	0·9	0·5	0·3
Threonine	1·6	0·9	0·4
Glutamic acid	2·8	1·4	0·6
Proline	0·3	0·5	0·2
Glycine	1·3	0·9	0·4
Alanine	1·1	0·8	0·5
β-alanine	—	—	0·1
γ-aminobutyric acid	—	—	0·1
Valine	1·1	0·7	0·3
Isoleucine	0·7	0·4	0·2
Leucine	1·3	0·6	0·2
Tyrosine	0·4	0·2	0·1
Lysine	—	—	0·3
Histidine	—	—	0·1
Arginine	—	—	0·1
Phenylalanine	0·4	0·3	0·1
Cysteine	0·1	0·1	—
Methionine	0·1	0·1	—
% nitrogen in soil	0·5	0·4	0·3

of these amino acids in soils hydrolysed with hydrochloric acid. Putnam and Schmidt[383] investigated the free amino acids in soil and found quantities ranging from 2 to 387 mg per g of soil of aspartic acid, glutamic acid, valine and leucine. Later Schmidt, Putnam and Paul[435] followed the disappearance of amino acids added to soil by measurement of carbon dioxide evolution and chromatographic analyses. They found that most of the amino acids were degraded in the first 48 hours (Fig. 16) and that they had almost entirely disappeared after 96 hours. Not all amino acids are so susceptible to attack, lysine and tyrosine being amongst the most resistant[165, 383]. In aerobic conditions, amino acids are broken down to ammonia and carbon dioxide but in anaerobic environments, large quantities of volatile fatty acids and ammonia are formed.

Another group of relatively simple substances found in soil are the leaf pigments. The presence of low concentrations of chlorophyll and carotenoid derivatives in the organic matter on the surface of woodland

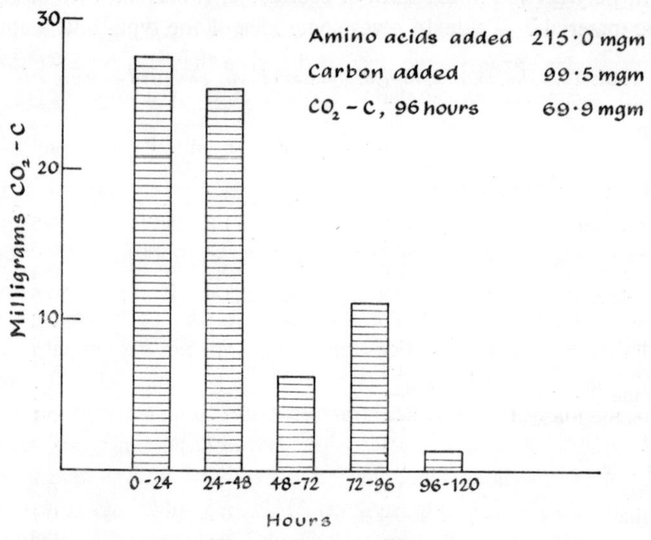

Fig. 16. *The disappearance of amino acids from soil, as measured by the evolution of carbon dioxide (Schmidt, Putnam and Paul*[435]*)*

(Reproduced by permission of E. L. Schmidt, E. A. Paul and the Soil Science Society of America)

soils was demonstrated by Gorham [149] and Gorham and Sanger[150]. In mor litter, the amount of these substances was twice that in mull; their concentration seemed to increase as soil pH dropped. Before leaf fall, chlorophyll broke down faster than the carotenoids, but after leaf fall the reverse was true. The breakdown of ^{14}C-labelled chlorophyll *b* and carotene was studied by Mayaudon and Simonart[313]. These substances decomposed much more slowly than glucose, hemicellulose or cellulose. Hoyt[217, 218] found that in addition to the supply from falling plant remains, some chlorophyll also arrived at the soil surface in the faeces of herbivorous animals. He observed that when chopped rye grass leaves were added to soil, 90 per cent of the chlorophyll was decomposed quickly. This was brought about by enzymes in the plant tissue. After this rapid initial loss, decomposition by soil micro-organisms was much slower; breakdown of both chlorophyll *a* and *b* took from 2 to 4 months.

Decomposition of Complex (Polymerised) Organic Compounds

The components of organic matter with a relatively complex molecular structure, e.g. polymers, are often insoluble and must be broken down by extracellular enzymes and absorbed before they can be assimilated by micro-organisms. We shall now consider some of the more important substances of this type.

(*a*) *Starch* (Fig. 17a)

This is the most common reserve foodstuff of plants, and although starches in different plants vary in their physical properties, chemically they are all very similar. Starch is a mixture of two polymers of glucose, amylose and amylopectin. Amylose consists of a chain of glucose units joined by α 1-4 glucosidic links. Amylopectin has the same type of chain but with side-chains attached by α 1-6 links. In laboratory conditions, many bacteria, actinomycetes and fungi isolated from soil will hydrolyse starch by producing extracellular enzymes known collectively as amylases. α amylase reduces both amylose and amylopectin to units consisting of several glucose molecules (dextrins). β amylase reduces amylose to maltose and amylopectin to a mixture of maltose and dextrins. Maltose is finally hydrolysed to glucose by glucosidase. The α 1-6 links of amylopectin are not broken by any of these enzymes and their cleavage is brought about by other enzymes. Surprisingly, the utilisation of starch in soil has received little study. However, Chalvignac[74] pointed out that only a few soil organisms were able to degrade starch to organic acids and carbon dioxide; most of them could only convert it to dextrins. He concluded that the breakdown of starch in soil took place in two phases; firstly, conversion to dextrins, and secondly, fermentation of these dextrins to simpler substances. Later de Barjac and Chalvignac[20] showed that breakdown of starch occurred most rapidly in fertile soils, less rapidly in infertile soils, and slowest of all in peaty soils. Prévot and Pochon[379] found that in dry and open soils, aerobes and facultative anaerobes predominated; if soil was compacted and water saturated, then obligate anaerobes such as *Clostridium butyricum* and *Clostridium amylolyticum* took their places.

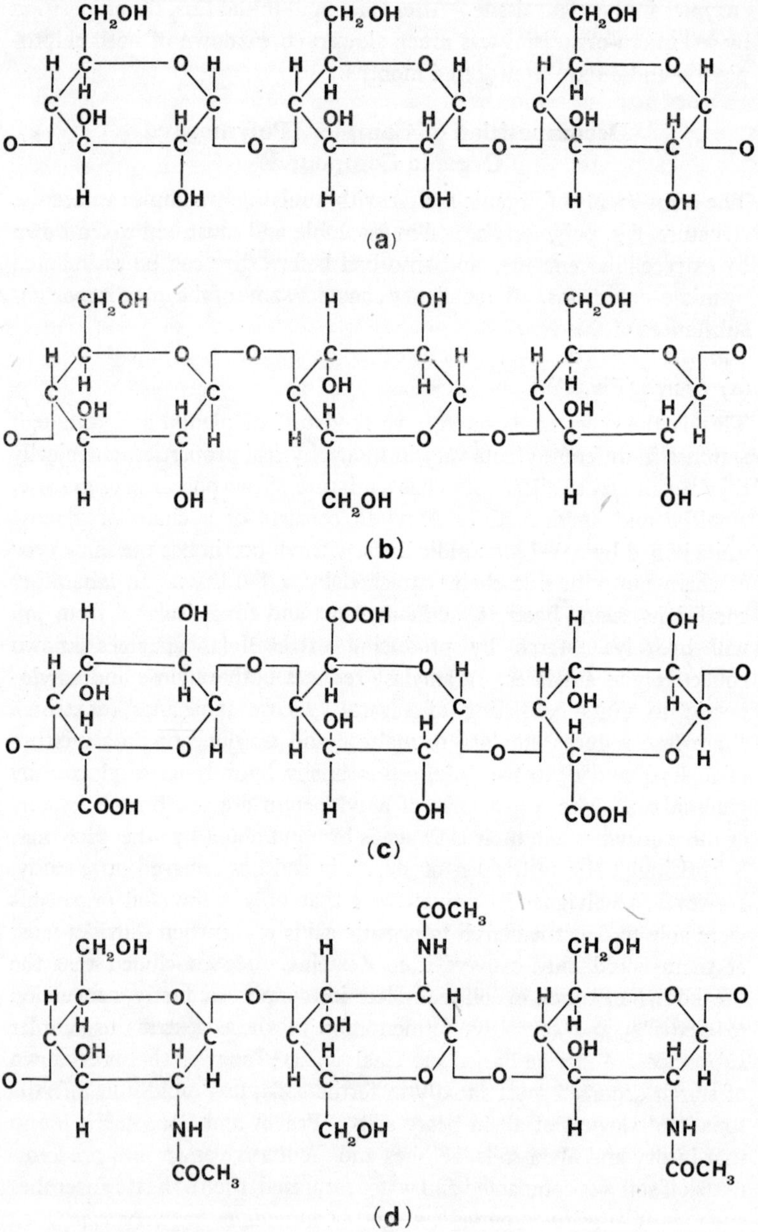

(a)

(b)

(c)

(d)

Fig. 17. (a) *Starch.* (b) *Cellulose.* (c) *Pectin.* (d) *Chitin* (*b*) *Cellulose* (Fig. 17b)

We noted earlier that cellulose was one of the most important constituents of plant tissues where it is often associated with other substances such as hemicelluloses, lignin and cutin. It also occurs in the cell walls of some fungi and the capsular material of *Acetobacter*. However, analyses of soil show that cellulose represents only a very small fraction of the organic matter, often less than 1 per cent[179], presumably because it is readily utilised by soil micro-organisms.

Cellulose is composed of glucose units bound in a long chain by β 1-4 linkages, the number of units involved varying in different plants. It is hydrolysed by a complex of enzymes, not yet fully characterised, given the collective name cellulase. Cellulases are possessed by a variety of bacteria, fungi and actinomycetes, including species of *Pseudomonas*, *Achromobacter*, *Bacillus*, *Clostridium*, *Streptomyces*, *Cytophaga*, *Trichoderma*, *Chaetomium* and *Coprinus*. The products of decomposition vary with the type of organism involved and the environment; in aerobic conditions cellulose is broken down to glucose, which may be incorporated in growing cells, and carbon dioxide; anaerobic bacteria convert it to various organic acids and alcohols.

Cellulose decomposition is a subject amenable to study since cellulose can be obtained in pure powder form for incorporation into culture media and as a film or paper for burial in soil. The precise form of the cellulose, e.g. cellophane or filter paper, can influence the development of cellulolytic organisms (see also p. 58), but the presence of impurities can also have important effects. Basu and Ghose[25] found that production of cellulase by fungi was stimulated by the presence of hemicelluloses. Not unnaturally, cellulose decomposition is also affected by the environment. Griffiths and Jones[176], who studied the effects of factors such as moisture content, nutrient supply and pH on colonisation of cellulose film in soil, found that when lime was added and the pH rose, the incidence of *Trichoderma viride* on the film dropped sharply, while the addition of phosphate stimulated the growth of *Streptomyces* spp.

Much information on the microbial colonisation of cellulose film in soil has been provided by the work of Tribe[514]. He studied the colonisation of film in various Canadian soils and observed a general pattern of succession. The initial colonisers were fungi, the dominant forms being *Rhizoctonia solani*, *Humicola* sp. and *Botrychium* sp. The latter two fungi formed special 'rooting hyphae' which grew into the film. Bacteria and nematodes appeared next, partly because the dead hyphae of the initial colonists provided them with a substrate. Finally, larger soil animals such as mites and collembola appeared. Later, Tribe[517] attempted to analyse the factors leading to the dominance of particular fungi by inoculating pairs of known cellulolytic fungi on to film and observing what happened. He concluded that the degree of cellulolytic activity exhibited by a fungus was only secondary to other competitive characteristics in allowing it to get established. Thus *Stachybotrys atra* was more competitive than *Rhizoctonia solani*, although the latter was the more strongly cellulolytic. The complete incorporation of cellulose film into humus has also been followed by Tribe[516]. He found that after 40 weeks in the soil, the humus formed from cellulose consisted almost entirely of products of microbial synthesis and that 84 per cent of the cellulose carbon had been liberated as carbon dioxide. The speed with which this process occurred was much affected by the nitrogen content of the soil. Cellulose itself contains no nitrogen, so that in the early stages of decomposition, nitrate is removed from the soil by the growing organisms. Later, this is released as the cells die and decompose[518].

(*c*) *Hemicelluloses*

Hemicelluloses occur in the thickened walls of cells in plant stems, roots, leaves and seeds. They are of two distinct types, polyuronides which consist of repeating units of sugars and uronic acids, and cellulosans which consist solely of sugars. Chemically, they are not related to cellulose for they contain a variety of five carbon sugars, as well as six carbon sugars, e.g. xylose, arabinose, galactose and mannose. Among the commonest types are the pectins, pentosans, mannans, xylans and galactans.

Xylans consist of chains of β 1-4 linked D-xylopyranose units, with side-chains of glucuronic acid or arabinose. Like other hemicelluloses it is difficult to purify as it is closely associated with other chemicals; their separation often results in changes in chemical structure. An enzyme complex termed xylanase is responsible for its breakdown and is found in a variety of organisms including *Streptomyces*, *Clostridium*, *Cellvibrio* and some phycomycetes. It has been suggested that fungi are most active at the beginning of xylan decomposition but that actinomycetes, once they have started growth, can keep up a more uniform attack over a longer period[535]. Sørensen[459], using radioactive tracer techniques, found that 65 per cent of the carbon in xylan was released as carbon dioxide whereas in the same period only 4 per cent was released from lignin. Compared to other hemicelluloses (e.g. galactan), xylan and mannan are less resistant to decomposition[535].

The middle lamellae of plant cell walls contain another group of hemicelluloses, the pectic substances built up from β 1-4 linked galacturonic acid units (Fig. 17c). There are considerable differences between the many pectic substances however, chiefly in respect to their chain length and solubility. *Protopectin* is water insoluble, while *pectin* is water soluble and partly esterified and *pectic acid*, water soluble and not esterified. Three enzymes are concerned with the breakdown of these materials:

(*i*) Protopectinase, which converts protopectin to soluble pectin;
(*ii*) Pectin methyl esterase, which attacks the methyl ester links of pectin to give pectic acid and methanol;
(*iii*) Polygalacturonase, which attacks the links between the galacturonic acid units of both pectin and pectic acids.

Many soil micro-organisms, e.g. *Erwinia*, *Clostridium*, *Pseudomonas* and *Bacillus*, possess these enzymes and, as we shall see later (Chapter 6), they may be involved in the penetration of legume root hairs by rhizobia. A number of soil-borne fungal pathogens, which induce wilt diseases and decay of stored vegetables, also possess pectinases.

(*d*) *Lignins*

Cells with lignified cell walls are found in many plant tissues and lignin-like polymers are found in fungi such as *Humicola*, *Aspergillus* and *Gliocladium*[365]. Lignin occurs in association with cellulose, forming lignocellulose, and the proportion of lignin increases as thickening progresses. Purification of lignin is difficult because of its intimate association with cellulose.

The lignins are not a chemically uniform group of compounds and their structure in different plants may not be the same. Thus the oxidation of conifer lignin yields vanillin and p-hydroxybenzaldehyde, while lignin from woody dicotyledons gives vanillin and syringaldehyde Fig. 18a, b, c). Their basic structure is aromatic, consisting of a polymer of phenylpropane units, but aliphatic hydroxyl and carbonyl groups are also present[221].

Considerable amounts of lignified material reaches the soil from both the aerial parts of plants and their roots. Its decomposition is very slow compared with the substances mentioned so far. Mayaudon and Simonart[313] compared the decomposition of cellulose and lignin in soil and found that the cellulose decomposed almost three times as quickly. The numbers of soil micro-organisms capable of breaking down lignin is rather small. Experiments by Sørensen[458] indicated that aerobic, Gram-negative, non-sporing rods, e.g. *Pseudomonas* and *Flavobacterium*, could decompose lignin, while Waksman and

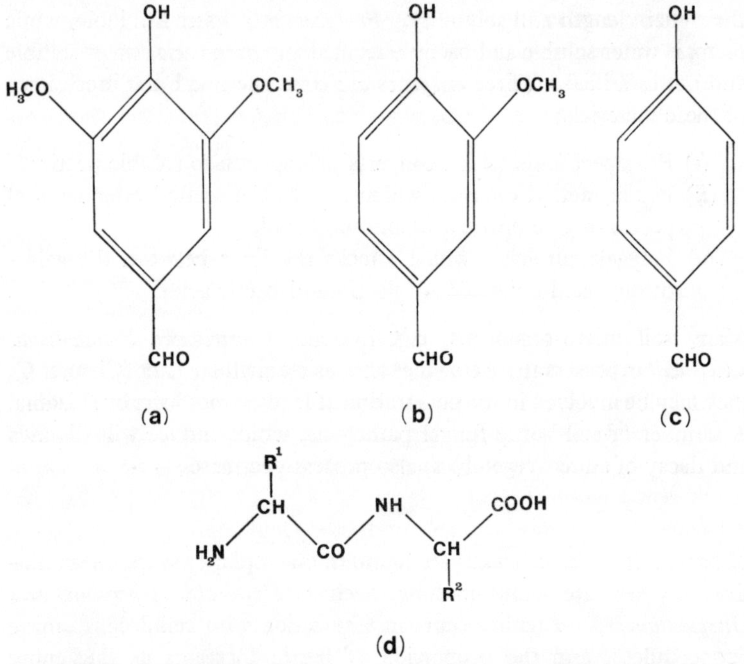

Fig. 18. (a) *Syringaldehyde.* (b) *Vanillin.* (c) *p-Hydroxybenzaldehyde.* (d) *Peptide linkage*

Hutchings[536] suggested that actinomycetes were also involved when present in mixed populations with other soil organisms. This latter idea has never been confirmed. Undoubtedly, the most important agents in the breakdown of lignin in the soil are the fungi, particularly basidiomycetes and ascomycetes which are the best suited for the decomposition of hard, woody substrates. Their hyphae, often aggregated into strands and rhizomorphs, are capable of penetrating into these materials for considerable distances by a combination of the physical forces of growth and secretion of enzymes[143]. In fact, the best known lignin-decomposing fungi are not strictly soil inhabitants at all, but are wood-rotting forms found on decaying logs above the soil surface, e.g. *Polystictus versicolor*, a common bracket fungus. Indirect evidence of the ligninolytic activity of true soil fungi has been obtained by Henderson[203, 204], who enriched soil with aromatic compounds related to lignin. She incorporated these compounds in kaolin pellets to provide localised substrates and isolated several fungi that could utilise them, e.g. *Pullularia pullalans*, *Trichosporum cutaneum*, *Phialophora aurantiaca* and *Margarinomyces* spp. Jones and Farmer[240], using a similar approach, found a number of fungi, including *Stilbum* spp. and *Humicola* spp., which could utilise vanillic acid and other breakdown products of lignin as sole carbon sources. Also there was some evidence for utilisation of lignin itself in cultures of *Humicola*, *Arthrobotrys* and *Cephalosporium*.

(*e*) *Lipids*

Many plant and animal cells can produce a variety of lipids, including fats and waxes. Fats are complex esters of fatty acids and glycerol, while waxes are esters of fatty acids with higher monohydric alcohols. They are broken down by lipases (esterases) to their constituent acids and alcohols, e.g. tributyrin is degraded to form butyric acid and glycerol. A variety of micro-organisms, especially bacteria, are known to attack fats and waxes[72], but comparatively little is known about their breakdown in the soil.

Some information is available on the breakdown of the surface lipids of plants, e.g. cutin (Fig. 19). Cutin is secreted from the epidermal cells of the plant and hardens on oxidation by atmospheric oxygen. Often, it is coated with a thin layer of waxy substances, constituting the so-called bloom of plants. Cutin is often formed as a layer above the pectic middle lamella of the cell but it may also be intimately mixed with cellulose (Fig. 20). In either case, it protects the cellulose and pectin from microbial action and must be removed before these substances can be attacked.

The chemical constitution of cutin is uncertain but it is thought that each molecule contains 16-18 fatty acids, linked together by ester linkages and occasional oxygen bridges. Two enzymes are involved in its breakdown, cutin esterase, which attacks the ester linkages and carboxycutin peroxidase, which splits the oxygen bridges[200].

Cutin may be partially decomposed before it reaches the soil by organisms present on the leaf surface (phyllosphere); Ruinen[422] showed that yeasts and *Azotobacter* could attack the cuticle and thus, increase leakage of nutrients from the leaf; Heinen and de Vries[202] have also shown that amongst the commonest soil organisms to attack cutin are the yeast *Rhodotorula* and the fungus *Penicillium spinulosum;* Gray and Lowe[158], who studied the breakdown of strips of purified

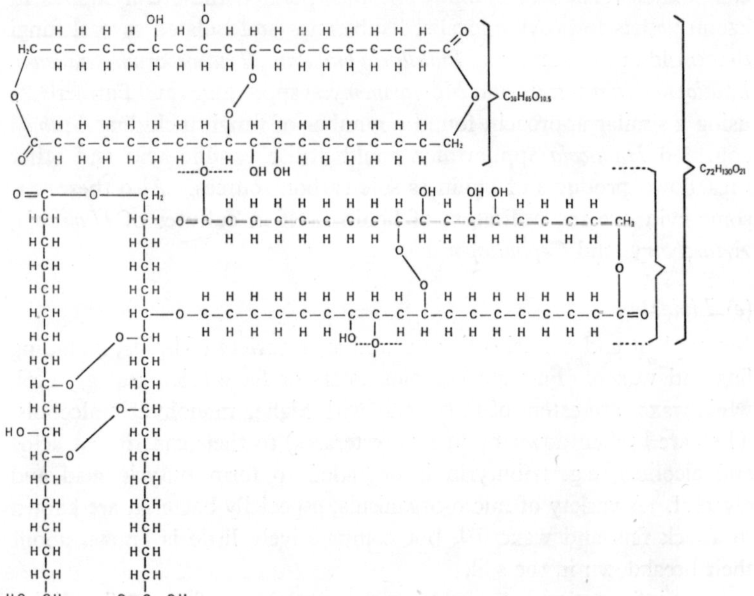

Fig. 19. *Cutin* (*Heinen*[201])
(*Reproduced by permission of* Zeitschrift für Naturforschung)

cutin buried in a pine forest soil, found that *Bacillus subtilis*, *Mortierella marburgensis* and a *Streptomyces* sp. could all break down cutin.

(*f*) *Chitin* (Fig. 17d)

Chitin is an important component of arthropod exoskeletons, fungal cell walls, some algae and nematode eggs. Like cellulose, it is made up of long-chain molecules but these consist of glucosamine rather than glucose. This means that chitin breakdown is unlikely to be limited by lack of nitrogen since nitrogen occurs in glucosamine. The products of decomposition are glucose and ammonia, both of which are readily available to most micro-organisms. The effect of this built-in nitrogen supply is obvious if the colonisation of chitin and cellophane films in soil is compared. Cellophane films often have large patches which remain undecomposed for long periods of time, while chitin is colonised

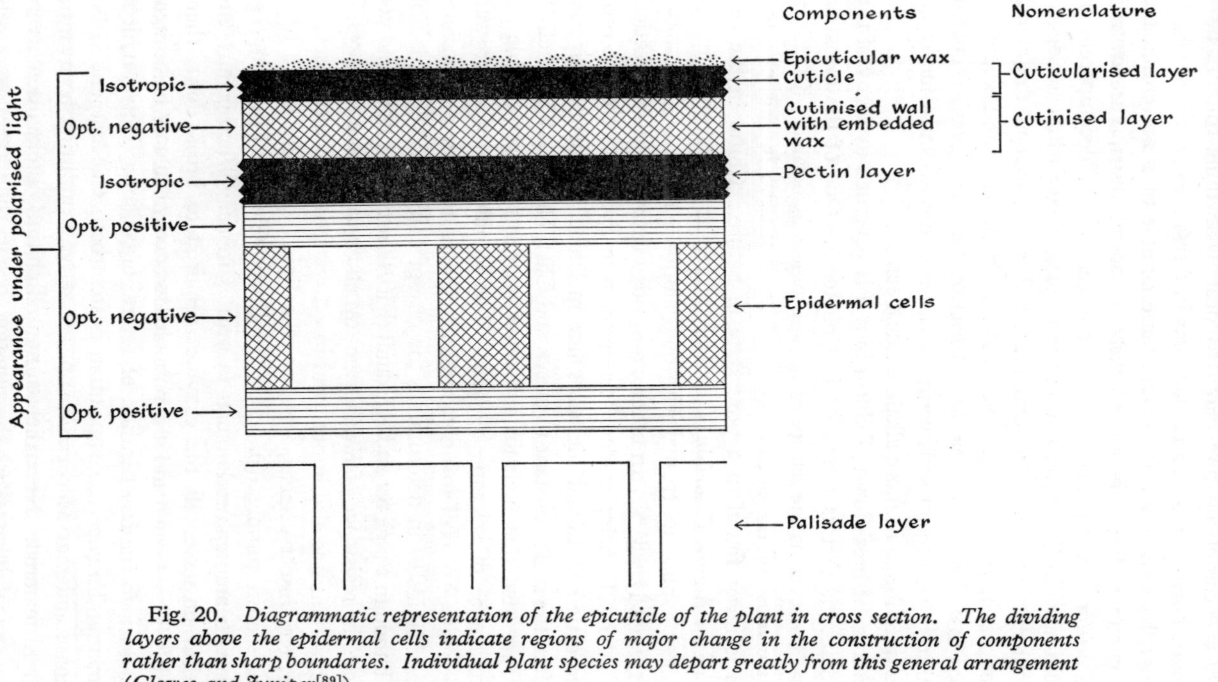

Fig. 20. *Diagrammatic representation of the epicuticle of the plant in cross section. The dividing layers above the epidermal cells indicate regions of major change in the construction of components rather than sharp boundaries. Individual plant species may depart greatly from this general arrangement (Clowes and Juniper*[89]*)*

(Reproduced by permission of B. E. Juniper and Blackwell Scientific Publications Ltd.)

rapidly and completely and may disappear in anything from 12 days in aquatic environments to 12-32 weeks in forest soils.

Fungi, bacteria and actinomycetes are all able to attack chitin, the most common forms being *Mortierella*, *Streptomyces*, *Pseudomonas* and *Bacillus*. Actinomycetes are prominent chitin decomposers in agricultural soils[525], especially in the tropics, bacteria in waterlogged environments and fungi in temperate soils[341], although the actual species present may be influenced by soil reaction. Gray and Baxby[155] showed that *Trichoderma viride* and *Mortierella marburgensis* occurred on chitin in acid forest soils while *Paecilomyces carneus* and *Mortierella alpina* were found in more alkaline forest soils.

The rate of breakdown of chitin in soil is governed by many of the environmental factors discussed in Chapter 2. One of the principal factors, however, is the nature of the substances associated with chitin, for, like most of the substances we have discussed in this chapter, chitin is never found in a pure state in nature. Chitin in the cell walls of the fungus *Fusarium* is probably protected by a layer of glucan, for unless hyphae are treated with both chitinase and β 1-3 glucanase, they are not lysed[454]. In other cases, this mixture is still ineffective, due to the walls being made of cellulose, as in *Pythium*, or to the protective action of fungal pigments like melanin. Pigmented mycelia are often the most resistant to attack and Chet, Henis and Mitchell[82] have suggested that melanin and other pigments are responsible for the resistance of sclerotia to lysis. The significance of other wall components, e.g. proteins, glycoproteins, mannans, other polysaccharides and lipids[15] in determining the rate of chitin breakdown is not clear although Potgieter and Alexander[375] showed that lipids did not increase the resistance of *Rhizoctonia* mycelium to enzyme mixtures.

Animal chitin is also impure and insect wings may consist of several layers, e.g. insect cement, wax, paraffin lipids, cuticulin and a lipoprotein chitin complex (p. 83). The lipoproteins are often impregnated with aminophenols, such as orthoquinone. Whether any one organism possesses all the enzymes required to break down these materials is not known and it is possible that decay can only be brought about through the co-operation of several organisms. However, it is known that *Streptomyces* spp. often possess both chitinase and β 1-3 glucanase and that *Mortierella* spp. are active in splitting both chitin and lipid material. Nevertheless, natural chitin is broken down very slowly and Veldkamp[525] showed that beetle wing cases were still intact after 6 months' burial in soils where pure chitin was broken down rapidly. Similarly, Okafor[342] found that insect wings could be recovered from soil after being buried for ten months (see Chapter 4).

(*g*) *Proteins*

Proteins represent the most abundant of the nitrogen-containing constituents of living organisms. They are readily attacked by many soil micro-organisms through the action of extracellular enzymes. Proteolytic enzymes hydrolyse the peptide links (Fig. 18d) between the amino-acid units that make up the protein, producing long chain polypeptides which are further hydrolysed to yield free amino acids. The breakdown of proteins, like gelatin and casein, in the laboratory is well known as these substances are often used in tests for identifying soil micro-organisms.

In the natural environment, proteins are rather more resistant to attack than they are in the laboratory. They may be associated with tannins in plant tissue[24] or with lignin. Estermann *et al.*[129] found that the digestion of ligno-protein complexes was very slow. However, soluble proteins are easily adsorbed by clay minerals such as kaolinite, and it has been suggested that since adsorption of both enzyme and substrate will take place, release of amino acids and ammonia will be enhanced. They showed that this took place when proteins were exposed to enzymes from *Bacillus subtilis*, *Pseudomonas* sp. and *Flavobacterium* sp.

There are also a number of insoluble, fibrous proteins, e.g. keratin, which find their way into soil. Keratin is resistant to attack by most proteolytic enzymes and is the principal component of skin, feathers, hair and claws. The integrity of the substance is due to the presence of disulphide bonds between the many cystine molecules present in it, but if these bonds are broken by autoclaving, ball-milling or microbial action, then the molecule is susceptible to attack by many proteolytic enzymes. The disintegration of autoclaved hair by a micro-organism does not prove that the organism produces a keratinase or an enzyme that splits disulphide bonds, or that it can break down native keratin in the soil However, Griffin[166] has established a pattern of colonisation of hair. The microbial succession culminated in the occurrence of keratinophilic fungi, including species of *Keratinomyces*, *Trichophyton* and *Microsporum* related to the causal organisms of ringworm disease. Griffin also noted the occurrence of *Streptomyces* species, supporting the observations made by Noval and Nickerson[335] on *S. fradiae* strains, isolated from keratin enriched soil, which could degrade keratin in wool sterilised with ethylene oxide. They also showed that *S. griseolatus* and *Nocardia polychromogenes* brought about the same reaction, resulting in the accumulation of soluble sulphydryl compounds (cysteine peptides). This suggests that the breakdown of keratin by microbes and insects is brought about by similar mechanisms for Waterhouse[551] showed that the guts of keratin digesting insects contained sulphydryl compounds.

Although keratin is not a common substrate in soil, keratinophilic organisms appear to be widespread. Pugh and Mathison[382] have suggested that they grow on other proteins when keratin is absent, for they found that *Arthroderma curreyi* and *Ctenomyces serratus* could grow and sporulate on dogfish egg cases in the laboratory. These structures are rich in chitin and protein, but as far as is known *Arthroderma* and *Ctenomyces* do not use chitin.

The comparative rarity of keratin in natural soils was also emphasised by other experiments carried out by Pugh and Mathison[382]. They isolated the imperfect conidial stage of *Ctenomyces* most frequently from open dune soils, while they found the perfect stage mostly in similar soils under bird traps where feathers were deposited. They suggested that since *Ctenomyces* is heterothallic (see p. 21), there would be little opportunity for strains of opposite mating type to meet in keratin-poor habitats, accounting for the predominance of conidial imperfect forms in the open dune soils.

Formation and Decomposition of Humus

We have already considered briefly the composition of humus and emphasised its heterogeneity in Chapter 2. Here we are concerned with its formation and ultimate breakdown. The decomposition of all the various substrates from plants, animals and microbes, by zymogenous micro-organisms, leads eventually to the formation of humus, which is then decomposed very slowly.

It is natural, but incorrect, to equate humus with the resistant remains of the original organic matter. It is partly a product of synthesis by the various organisms involved in the decomposition processes. Various studies with radioactive tracers have shown that many chemical substances can contribute to the formation of humus. After 7 days in soil, 70-75 per cent of glucose ^{14}C was located in humic fractions[310] and acetate ^{14}C was widely distributed after 6-9 hours in soil amongst the humin, fulvic acid and humic acid fractions[480].

^{14}Carbon, in decomposing barley straw, was found uniformly distributed between these fractions[459]. Tracer work has indicated the importance of lignin in the formation of humus and Mayaudon and Simonart[313] found that 34·2 per cent of the radioactivity in humic acid originated from lignin while only 5·9 per cent came from cellulose. The cellulose carbon contributed mainly to the simpler components of humus, e.g. amino acids, while the lignin carbon was found in more complex substances, possibly phenolic compounds. Since lignin decomposes more slowly than many other substances, it makes up an increasing percentage of residual soil organic matter.

The aromatic core of humic acid is probably formed by the polymerisation of polyphenolic units to which the simpler compounds like amino acids may be attached. The aromatic compounds may originate from plant material or be produced by microbial synthesis. Burges, Hurst and Walkden[61] showed that in some instances relatively unchanged lignin residues, e.g. vanillic and syringic acids, could be detected when degradation products of humic acid were studied by thin layer chromatography. In soils where lignin was absent, e.g. moss-covered Antarctic soils, these residues were not found in humic acid and other degradation products, including flavonoids, were detected which could have been derived from microbial synthesis or the vegetation (Table 17).

The decomposition of humus is an extremely slow process and it has been shown by carbon-dating techniques that some humic fractions in soil are very old. Paul *et al.*[358] estimated that the humin and humic acid fractions from a chernozem soil had a mean residence time of over 1,000 years. They also showed that cultivation of soil could lead to a reduction in the mean residence time from 2,200 years to 500 years, indicating an increased turnover of soil organic matter. Alexander[8] has suggested that the resistance of humic compounds to decomposition is due to several factors, including the presence of lignin derivatives complexed with proteins, the polyphenolic nature of the molecule, the ready adsorption of its carbonaceous components by silicate minerals and the formation of complexes with trivalent cations. Swaby and Ladd[499] also suggested that resistance was due to the molecules containing diverse building blocks irregularly cross-linked by a variety of covalent bonds: degradation would only be possible if many extracellular enzymes from many different organisms were formed.

A number of difficulties are encountered when investigating the microbial decomposition of humic acid. Firstly, microbes may grow

Table 17. *Degradation products of humic acids from different soils*[221]

	Podzol	Humic acid (L. Light & Co.)	Rendzina	Chernozem	Mull humus (oak)	Antarctic moss
POSSIBLE FLAVONOID UNITS						
Phloroglucinol	+++	+++	Trace	++	–	++
Resorcinol	+++	+++	+	+++	++	++
o-methyl phloroglucinol	++	++	Trace	++	–	+
2-4 dihydroxytoluene	++	++	Trace	+	–	+
Pyrogallol	+	+	+	+	–	–
3-5 dihydroxybenzoic acid	++	+	Trace	+	–	+
POSSIBLE LIGNIN UNITS						
p-hydroxybenzoic acid	+	+	++	+	++	–
Vanillic acid	++	++	+++	++	++	–
Syringic acid	–	–	++	+	++	–
Protocatechuic acid	+	+	+	+	+	–
Guaiacylpropionic acid	+	–	++	+	–	–
Syringylpropionic acid	–	–	+	+	–	–

on humic acid media but use only associated compounds such as amino acids leaving the aromatic core intact; secondly, lack of agreement on the exact structure of the molecule (if there is a common basic molecule) makes comparison of different studies difficult. Finally, as Burges and Latter[62] showed, zones of clearing around colonies growing on media containing humic acid does not necessarily indicate decomposition of humic acid; in some cases humic acid is precipitated in a sheath around the microbial cells. Nevertheless, it seems that certain fungi, particularly ascomycetes and basidiomycetes, can decompose humic acid in laboratory conditions by enzymatically reducing aromatic carboxyl groups to aldehydes and then alcohol[222]. *Penicillium frequentans* was shown to use humic acid as a sole source of carbon, the aromatic moiety being used in preference to the aliphatic[308]. The enzyme systems operating were adaptive. Salicyl alcohol and salicylaldehyde were detected in culture filtrates of the fungus growing on humic acid[309]. As yet, we have little information on the decomposition of humic compounds in nature.

Decomposition of Pesticides

With an ever-increasing requirement for high crop yields, the application of pesticides to crops and soil has become commonplace. Much concern has been expressed about the effect of these pesticides on soil processes, crop plants and animals, including man, who may eat these crop plants. The extent of unwanted side-effects may well be governed by the activities of the soil microflora, which may activate or detoxify pesticides, or even convert pesticides into growth promoting substances.

In general, the most side-effects can be expected from persistent pesticides as these may accumulate in plant tissues over a long period of time, may be leached into public water supplies or may affect the growth of susceptible crops planted in subsequent years. Many of the pesticides that are available are persistent, chiefly because they provide prolonged control of pests and are cheaper to use. They include chlorinated hydrocarbon insecticides (aldrin, DDT), organophosphate insecticides (parathion), phenoxy herbicides (2,4,5-T), phenylurea herbicides (diuron), triazine herbicides (atrazine, simazine) and fungicides such as captan[8].

Although it is premature to say exactly what determines the persistence of these substances in soil, Alexander[9] has pointed out certain characteristics of pesticides which decompose slowly. In general they are aromatic substances which are partially substituted, so that chlorophenols are more persistent than unsubstituted phenols. Amino, methoxy, sulphonate and nitro substitutions all increase resistance, while hydroxy and carboxyl groups increase susceptibility. Resistance is also increased by increasing the number of substituents (diaminobenzene is more resistant than monaminobenzene) and by placing the substituents in a suitable position on the benzene ring (o- and p- chlorophenols are susceptible but m-chlorophenol is resistant). These results are summarized in Table 18.

Table 18. *Breakdown of substituted benzoic acids by the soil microflora*[9]

Substitution	Days for ring cleavage: ortho	meta	para	
$-COOH$	2	8	2	
$-OH$	2	2	1	Unsubstituted
$-NH_2$	2	>64	8	benzoic acid—
$-OCH_3$	4	16	2	1 day
$-SO_3H$	>32	>32	>32	
$-NO_2$	8	>32	4	

The reasons for these changes in resistance are not clear, and will probably remain so until the relative importance of the different methods of breaking-down these molecules are ascertained. These methods include the conversion of non-fatty acid side chains to fatty acids, the β-oxidation of long fatty acid moieties to short alkanoic acid moieties, cleavage of the ether link between the side-chain and the ring, ring hydroxylation and ring cleavage[248].

6

Effects of Living Plants on the Soil Microflora

Living roots provide a localised and continuous source of nutrients for the soil microflora in the form of live tissues, exudates and sloughed-off dead cells. They also provide a continuously expanding surface for colonisation and by their metabolic activities change the soil environment in their immediate vicinity. It is not surprising, therefore, that great changes in microbial activity occur in the root region which may have both beneficial and harmful effects on plants (see Chapter 7). Many microbes are only loosely associated with roots, occurring in high numbers in an ill-defined zone of the soil around the root, termed the rhizosphere, while some organisms can form a closer association by growing on the root surface (or rhizoplane). A smaller number of microbes can invade the living root cells where they live as symbionts or parasites. In this chapter, the importance of the root to these various groups of microbes will be considered.

The Root Environment

Various explanations for the enhanced activity of micro-organisms around roots have been suggested.

It is possible that the solid surface which the root provides can itself increase the amount of growth, particularly of filamentous forms such as fungi. This possibility was tested by Parkinson and Pearson[351], who compared the development of fungi on live and dead roots with that on inert nylon threads, of similar dimension to the roots, buried in the same soil. The results indicated that the surface provided by the threads had little or no effect on soil fungi. The possibility that root surface inhabitants do gain some advantages by having a more or less continuous surface for growth cannot be entirely ruled out. Indeed,

Brown, Jackson and Burlingham[55] have suggested that the presence of a continuous root surface is one of the reasons for the spread of *Azotobacter* inoculated into soil.

It is clear that the metabolic activities of roots have a marked effect on the soil environment around them. Certain gases and ions are taken in from the soil, while others are passed out. Many such exchanges occur and it is therefore difficult to assess their overall influence on the soil environment. It is often assumed, for example, that roots increase acidity in soil around them by excretion of carbon dioxide and also possibly H^+ ions. However, it has been suggested that roots take up on average more anions than cations and would therefore tend to pass out HCO_3^- ions, rather than H^+ ions, to preserve electrical neutrality[339]. This would increase the pH of soil around the roots and possibly counteract the effect of carbon dioxide which would diffuse away from the root region more rapidly than the bicarbonate ions. Nye suggested that, in an acid soil, a rise of one pH unit in soil a few millimetres around the root was possible. Such an increase would enhance the activity of many soil micro-organisms, although by itself cannot account for all cases of stimulation.

The most widely accepted explanation for the stimulation of soil micro-organisms is that the root provides a supply of nutrients. Pathogenic and symbiotic microbes clearly obtain this from living root cells, either by dissolving the cell walls and penetrating into the cytoplasm or by growing between cells and obtaining nutrients by diffusion through the cell walls. Micro-organisms on and around the root are provided with a number of possible nutrient sources. It is known that exudation of organic substances from roots occurs, brought about by active processes, passive diffusion or exudation from moribund cells[66]. Also surface cells of roots are continually being sloughed-off into the soil. In young roots, these are mainly root hairs and root cap cells but with the onset of secondary thickening, many of the outer cortical layers are detached. The surface of young roots is often covered with a layer of mucilage (the 'mucigel') and electron micrographs have shown that microbes become embedded in this[232] (Plate IV (d)); this too may provide a source of nutrients for at least some of the root surface microflora.

The work of Rovira[411-413] provided detailed evidence of the importance of root exudates to soil micro-organisms and subsequently a large number of compounds have been detected in exudates from a variety of plants. These have usually been detected by paper chromatography of exudates from plants grown in aseptic conditions, often in sterile sand. According to Rovira[416], compounds found so far include 10 sugars, 23 amino acids, 10 vitamins, mucilaginous oligosaccharide material, organic acids, nucleotides, flavonones and a host of miscellaneous substances. Recently Boulter, Jeremy and Wilding[41] used the more sensitive technique of ion exchange chromatography to analyse exudates from peas and found a wider range of amino acids and peptide material. However, the composition of exudates is dependent upon a number of things, including the type and age of the plant. Thus, Rovira[411] compared exudates from pea and oat plants 21 days after germination and found 22 amino compounds from pea but only 18 from oats, and in a subsequent study[415] showed that sugars were exuded only during the first few days of growth following germination. After the first few weeks, the total amount of exudation decreases and it seems likely that the more mature parts of root provide more nutrients in the form of sloughed-off cells. As one might expect, many of the substances detected in root exudates have a marked effect on the growth of micro-organisms in laboratory culture although less work has been done on the responses of populations in the soil. However, Kunc and Macura[258] added amino acids, organic acids and sugars to soil and after 76 hours, 85 per cent of the carbon had been liberated as carbon dioxide.

One of the main difficulties in assessing the importance of exudates is the lack of information on their production in the natural habitat. Results obtained with sterile roots are not necessarily applicable to those growing in soil, for the mere presence of micro-organisms around the root may alter the nature of the exudates. McLaren[299] has also found that the thickness of the mucigel around roots decreases under sterile conditions. Furthermore, the amino acid content of a root system growing under sterile conditions is different from that under non-sterile conditions, some amino acids, e.g. lysine, being more abundant in non-sterile roots. Boulter *et al.*[41] showed that even in aseptic conditions the nature of exudates was influenced by methods used to culture roots. However, it is reasonable to assume that exudation does occur in natural conditions, even though the composition of the exudates may be different. Use of radioactive tracers should shed more light on this problem, indeed preliminary studies with plants growing in an atmosphere containing $^{14}CO_2$ have shown that radioactive carbon is liberated into soil by roots[296].

Rhizosphere and Root-surface Microflora

The sphere of influence of the root in soil is difficult to define and for most practical purposes the rhizosphere is recognised by the increases in numbers of microbes in soil near to or remaining attached to roots when they are carefully dug up. All the main groups of micro-organisms occur in higher numbers in the rhizosphere than in soil distant from roots and the increases are expressed as an R: S ratio (rhizosphere: soil ratio). The highest ratios obtained are for bacteria, with lower ones for actinomycetes and fungi and lower still for algae and protozoa (Table 19). The stimulation effect has been detected by several

Table 19. *Comparison of the numbers of various groups of organisms in the rhizosphere of spring wheat and in control soil*[408]

Organisms	Rhizosphere soil	Control soil	S.D.	Approximate R:S ratio
Bacteria	$1{,}200 \times 10^6$	53×10^6	**	23:1
Actinomycetes	46×10^6	7×10^6	**	7:1
Fungi	12×10^5	1×10^5	**	12:1
Protozoa	24×10^2	10×10^2	**	2:1
Algae	5×10^3	27×10^3	*	0·2:1
BACTERIAL GROUPS				
Ammonifiers	500×10^6	4×10^6	**	125:1
Gas-producing anaerobes	39×10^4	3×10^4	*	13:1
Anaerobes	12×10^6	6×10^6	*	2:1
Denitrifiers	126×10^6	1×10^5	**	1260:1
Aerobic cellulose decomposers	7×10^5	1×10^5	*	7:1
Anaerobic cellulose decomposers	9×10^3	3×10^3	NS	1:1
Spore formers	930×10^3	575×10^3	NS	1:1
'Radiobacter' types	17×10^6	1×10^4	**	1700:1
Azotobacter	$<1{,}000$	$<1{,}000$	NS	?

* Figures significantly different at 5 per cent probability.
** Figures significantly different at 1 per cent probability.
NS Not significantly different.

methods, including plate counts, direct counts and respiratory measurements[247, 468, 472].

Root surface inhabitants are regarded as those which remain on the roots after vigorous washing with sterile water has removed all adhering soil. Unfortunately, bacteria and actinomycetes are more easily removed than are fungal hyphae and so less is known about their distribution and importance. It is generally assumed that the maximum stimulation of microbes outside the root occurs at the root surface, but this is difficult to prove since comparisons between 'areas' of root surface and 'weights' of soil are meaningless.

(a) Bacteria

Although most bacteria are stimulated by roots, some are more responsive and outstrip their competitors in the race for the nutrients provided. It has been shown that generally Gram-negative, non-sporing rods are the most stimulated, whilst Gram-positive, non-sporing rods, cocci and spore-forming bacteria are least affected. Vágnerová, Macura and Čatská[519, 520] observed that the fastest growing Gram-negative rods were found on the root surface, while the fastest growing cocci and spore-formers occurred in the root-free soil, indicating that these bacteria were specifically adapted to different environments. Macura[290] pointed out that successful colonisation of the root does not depend solely on growth rate, other contributing factors being the capacity to use substrates present, response to the physical environment and antagonism. Indeed, all the factors contributing to an organism's competitive saprophytic ability (p. 86) are important.

There are many indications that the nutritional requirements of root surface and rhizosphere bacteria differ from those active in root-free soil. Lochhead and Thexton[276] showed that bacteria able to utilise inorganic nitrogen and amino acids were stimulated in the rhizosphere, whilst those requiring growth factors were stimulated to a much smaller degree. Other investigations have shown that ammonifying, sugar-fermenting, acid-producing, cellulose-decomposing, denititrifying, vitamin-producing, polysaccharide-producing and phosphate-dissolving bacteria are stimulated in the rhizosphere (Table 19). However, even though these micro-organisms are found in increased numbers in the rhizosphere, the percentage of the total population which they comprise may not increase. Thus, although increased numbers of dicalcium phosphate-dissolving bacteria occurred on the root surface and in the rhizosphere of oat, they were not preferentially stimulated[281].

Most investigations have shown little correlation between the physiological and taxonomic groups of bacteria in the rhizosphere. However, Rouatt and Katznelson[407] were able to show that *Pseudomonas* species were dominant in the rhizosphere, while *Arthrobacter* species were characteristic of root-free soil; it is known that *Arthrobacter* species have complex growth factor requirements while *Pseudomonas* species grow on relatively simple nitrogenous media. Furthermore, Chan and Katznelson[75] showed that the relative incidence of *Pseudomonas* species also increased on the root-surface where they could outgrow and inhibit *Arthrobacter* in the presence of extracts from soybean roots.

Another bacterium which seems to be physiologically adapted to life in the rhizosphere is *Azotobacter*, yet most investigations show that its relative incidence is not increased in the rhizosphere. Chan, Katznelson and Rouatt [76] have suggested that this is due to the antagonistic effects of other rhizosphere organisms or to unfavourable environmental factors. However, Brown, Jackson and Burlingham[55] have provided evidence that populations of *Azotobacter*, artificially introduced into soil, will survive and spread on the root surface but show little spread in root-free soil. The status of *Azotobacter* in soil will be considered in more detail in Chapter 7.

(*b*) *Actinomycetes*

Less attention has been paid to actinomycetes in the rhizosphere although they are quite numerous and active around roots (Plate VII (a)). Studies have often concentrated on detecting antagonistic actinomycetes producing antibiotics which inhibit root pathogens. For example, Řeháček[392] found that 22 per cent of actinomycetes isolated from the rhizosphere of cereals inhibited potential pathogenic fungi such as *Fusarium oxysporum* and *Alternaria solani*. The importance of such antagonistic effects in the field has yet to be accurately assessed.

The types of actinomycetes found in the root region are similar to those from root-free soil; usually *Streptomyces* and *Nocardia* species predominate. The physiological activities of actinomycetes from rhizospheres and non-rhizosphere soil of several plants were compared by Abraham and Herr[1], who found that the only significant difference was that numbers of starch hydrolysers were higher in rhizosphere of corn and soybean. High counts of pectinolytic actinomycetes were found in the rhizosphere of *Calotropsis gigantea* (a tropical shrub) by Agate and Bhat[5], while Venkatesan[527], studying actinomycetes in the rice rhizosphere, found that those needing amino acids and growth factors were relatively more frequent in soil around the roots.

(*c*) *Fungi*

Many studies of fungi in the rhizosphere have involved the use of a dilution plate technique for both qualitative and quantitative assessments. The information obtained has been largely floristic and little is known about the physiology and nutrition of these forms. During the past 10 years considerable attention has been paid to root-surface inhabiting fungi (Plate VII (b)) and many workers have been able to define a distinct flora, with *Fusarium* species, *Cylindrocarpon radicicola* and non-sporing isolates being among the dominant forms. However,

Table 20. *Fungi associated with dwarf bean roots showing the estimated relative importance of each species in each of four recognisable habitats*[504]

Fungus	Root surface	Cortex	Outer stele	Inner stele
Mucor spp.	+	+		
Mortierella vinacea	+	+		
Trichoderma viride	++	+		
Fusarium sambucinum	++	+		
Penicillium spp.	++	++		
Penicillium lilacinum	++	++		
Mortierella spp.	++	++	+	
Gliocladium spp.	+++	++	+	+
Fusarium oxysporum	+++++	+++++	+	+
Cylindrocarpon radicicola	+++	+++	+++++	++++
Sterile dark forms	+	+	++++	+++++
Varicosporium elodea		+	++	++
Sphaeropsidales spp.		+	+	
Botrytis cinerea		+	+	
Sterile hyaline forms			+	++

+ Usually present but only low frequency of isolation.
+++++ Dominant form with high frequency of isolation.

these fungi are not necessarily confined to the root surface. Taylor and Parkinson[504] studied colonisation of the root surface, cortex and stele of dwarf bean roots and found that some fungi confined to the surfaces of young roots were able to penetrate tissues of older roots, e.g. *Fusarium oxysporum*, and *Cylindrocarpon radicicola* (Table 20).

The origin of the initial colonisers of root surfaces has been investigated by several workers. Apparently fungi and bacteria present on seed coats do not play a major role in the colonisation of the roots produced after germination[362]. A similar conclusion was reached by Parkinson, Taylor and Pearson[352], who observed that young roots were first colonised by a random selection of soil fungi which, after a few days, gave way to a more restricted microflora of typical root-surface forms which persisted until senescence.

It is generally considered that root-surface inhabitants are uncommon in rhizosphere soil. This is probably due to the different methods used to detect fungi in these two environments. Most isolations of rhizosphere fungi are made with dilution plate techniques which are selective for heavily sporing forms, whereas isolation of root-surface fungi are achieved by washing procedures which remove many spores. Parkinson and Thomas[353], who used both dilution techniques and a washing technique, found that while the latter detected *Fusarium* species and other root-surface fungi in the rhizosphere soil, the dilution methods did not, suggesting that these fungi are rhizosphere inhabitants which compete successfully for the root-surface habitat.

(*d*) *Algae*

Little is known about the effects of plant roots on algae although positive stimulation of algae in the rhizosphere of tea plants was noted by Hadfield[182]. Shtina[446] found that while rye and potatoes also stimulated algae, lupin, clover and timothy grass did not. Since algae are mainly autotrophic, it is unlikely that these effects are due to the presence of major nutrients in the exudates. It is more likely that growth factors, antibiotics and changes in the physical environment are involved.

(*e*) *Factors affecting the Nature of Rhizosphere and Root-surface Microfloras*

The qualitative and quantitative nature of the microbial populations around roots are influenced by many factors. Starkey[468-472], in a series of important papers, showed that the rhizosphere effect depended upon the type of plant, age of plant, health and vigour of the plant, the position and type of root, and the soil type and environment.

(*i*) *Plant and Root Type* Plant roots stimulate or suppress micro-organisms to varying degrees, although few cases of inhibition have been conclusively proved. Among crop plants, legumes usually cause most stimulation and Rouatt and Katznelson[407] obtained highest rhizosphere counts with red clover, intermediate counts with flax and oat and lowest counts with wheat, barley and corn. Some roots may inhibit microbes, for Hill[208] showed that fewer cells of *Bacillus subtilis* occurred on the root surface of *Pinus nigra* than in the rhizosphere soil. However, it has not yet proved possible to relate any of these changes to differences in exudation patterns of the plants.

Even over the surface of a single root, changes in the microbial populations and the nutrient regime may be considerable. Pearson and Parkinson[359] showed that roots secreted most amino acids in the region of rapid cell elongation and Vágnerová[518] found that numbers of bacteria requiring amino acids were greater at the tip of the root than at the base. Furthermore, the density of bacteria, fungi and algae in the rhizosphere may decrease with increasing distance horizontally and vertically from the base of the stem[511].

(ii) Age and Vigour of the Plant The rhizosphere effect begins to increase after germination of seedlings, reaches a peak around the time of flowering and fruiting and then declines sharply as root senescence begins[469].

However, the effect of ageing may be complicated by the influence of environmental factors on plant growth. According to Rovira[414], high light intensity and temperature increase the rate of root exudation while a decrease of light intensity has been shown to decrease the rhizosphere stimulation of bacteria[406]. However, Peterson[363] was unable to detect any effect of different light intensities on fungi but reported that high temperatures led to a stimulation of certain fungi (especially non-sporing, hyaline forms), and decreased bacterial numbers.

The role of plant metabolism in controlling root-region populations is now arousing some interest since plant metabolism may be altered by foliar application of various substances, such as antibiotics, urea, phosphate and growth regulators. Vraný[531, 532] observed that applications of chloramphenicol and urea to leaves not only changed the nature of root exudates but also brought about increased growth of fungi while inhibiting bacteria. Metabolic differences, e.g. secretion of cyanides, may also account for differential rhizosphere effects in plants resistant to soil-borne pathogens and those of the same species which are susceptible.

(iii) Soil Type and Environment The rhizosphere effect is most pronounced in sandy soils and least obvious in clay and humus soils[4]. In desert conditions the root region may be the only place where microbes can flourish[426], and sand-dune plants cause a marked stimulation of the microflora, e.g. *Atriplex babingtonii* has an R:S ratio of approximately 1450:1[554].

Within any one soil, environmental changes seem to have less effect on the rhizosphere microflora than on the populations in soil away from roots. The rhizosphere is then, to some extent, a 'buffered' environment. This was demonstrated by Katznelson and Richardson[246] who found that although additions of dried blood to soil caused large pH changes, the root microflora was not greatly affected. However, additions of fertilisers and pesticides to soil can alter the rhizosphere population. Gunner *et al.*[178], studying the effects of the pesticide diazinon, found that it caused changes in the rhizosphere microflora, resulting eventually in the predominance of *Streptomyces* species.

Symbiotic Association of Micro-organisms with Plant Roots

Symbiosis literally means 'living together' and the term was originally applied to parasitic associations, in which one partner exploits the other, as well as to mutualistic associations, in which both partners benefit without harming each other. However, the term 'symbiosis' is now frequently used to describe mutualistic associations and it is in this sense that it is used here. In this chapter, the characteristics of symbiotic plant-microbe relationships will be described and the benefits derived by the microbial partners outlined. The benefits gained by the plants from the associations will be discussed in Chapter 7.

(*a*) *The Rhizobium-legume Symbiosis*

The existence of swollen nodules on the roots of legumes (Plate VII (c)) and the value of legumes in maintaining soil fertility have been known for a long time. Their importance is due to the presence in the nodules of bacteria, which are able to fix atmospheric nitrogen and hence improve the nitrogen status of both plant and soil (see Chapter 7).

The bacteria in the nodules belong to the genus *Rhizobium*. By living in the root, they gain nutrients for growth and are able to escape from the rigours of competition with the saprophytic soil microflora. Rhizobia are Gram-negative, non-sporing rods, closely related to the bacterium causing crown gall disease (*Agrobacterium tumefaciens*) and to harmless root-surface inhabitants (*Agrobacterium radiobacter*). The various species of *Rhizobium* are differentiated by their host preferences. Host plants are placed in 'cross-inoculation groups' which may be defined as 'a number of host species, any one of which will form nodules when inoculated with bacteria from nodules occurring on any other member of the group'. The bacteria causing nodulation of plants in a cross-inoculation group are placed in the same species. In practice, the species and cross-inoculation groups are not easily defined since occasional cross-infections between groups can occur. Only about 10 per cent of the Leguminosae have been investigated so

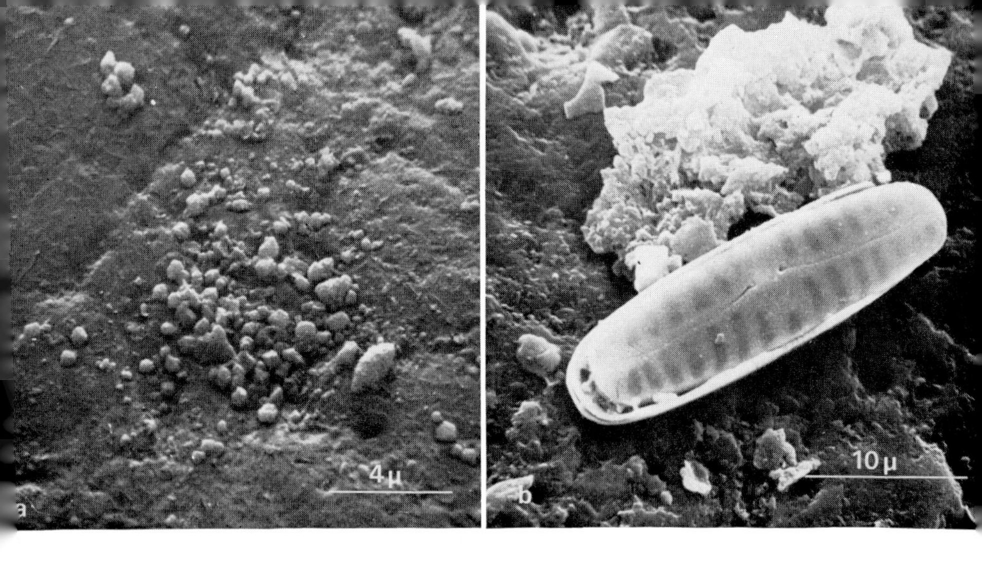

Plate I. *Growth patterns of micro-organisms on soil particle surfaces as seen with the scanning electron microscope.* (a) *Non-migratory, unicellular pattern of some bacteria,* (b) *migratory, unicellular pattern as exhibited by* Pinnularia borealis, *a diatom,* (c) *restricted hyphal pattern exhibited by a* Streptomyces *sp. on a dead root fragment,* (d) *mycelial strand pattern as exhibited by hyphae in a sand-dune soil*

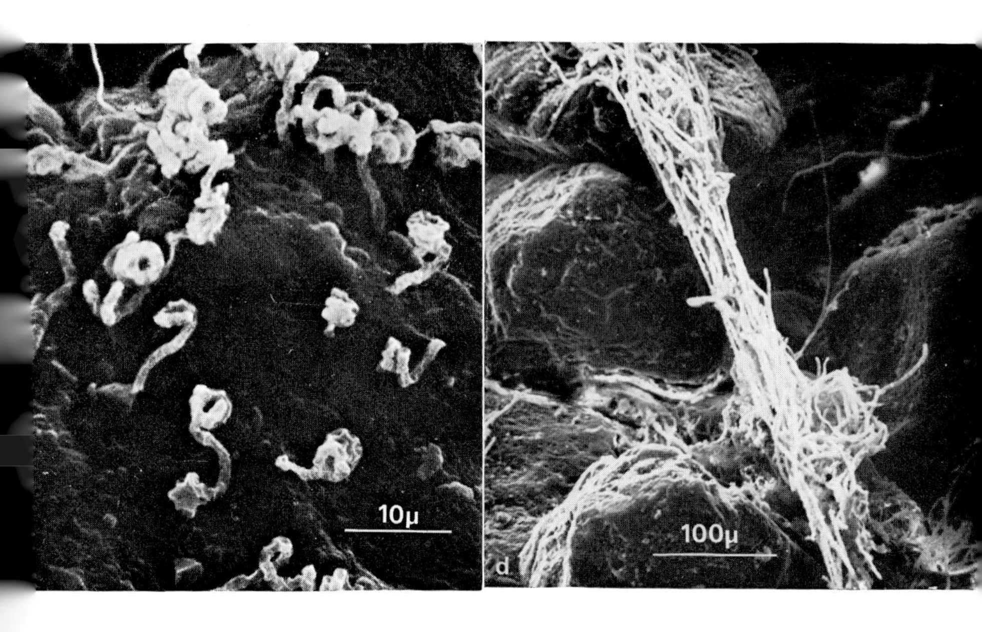

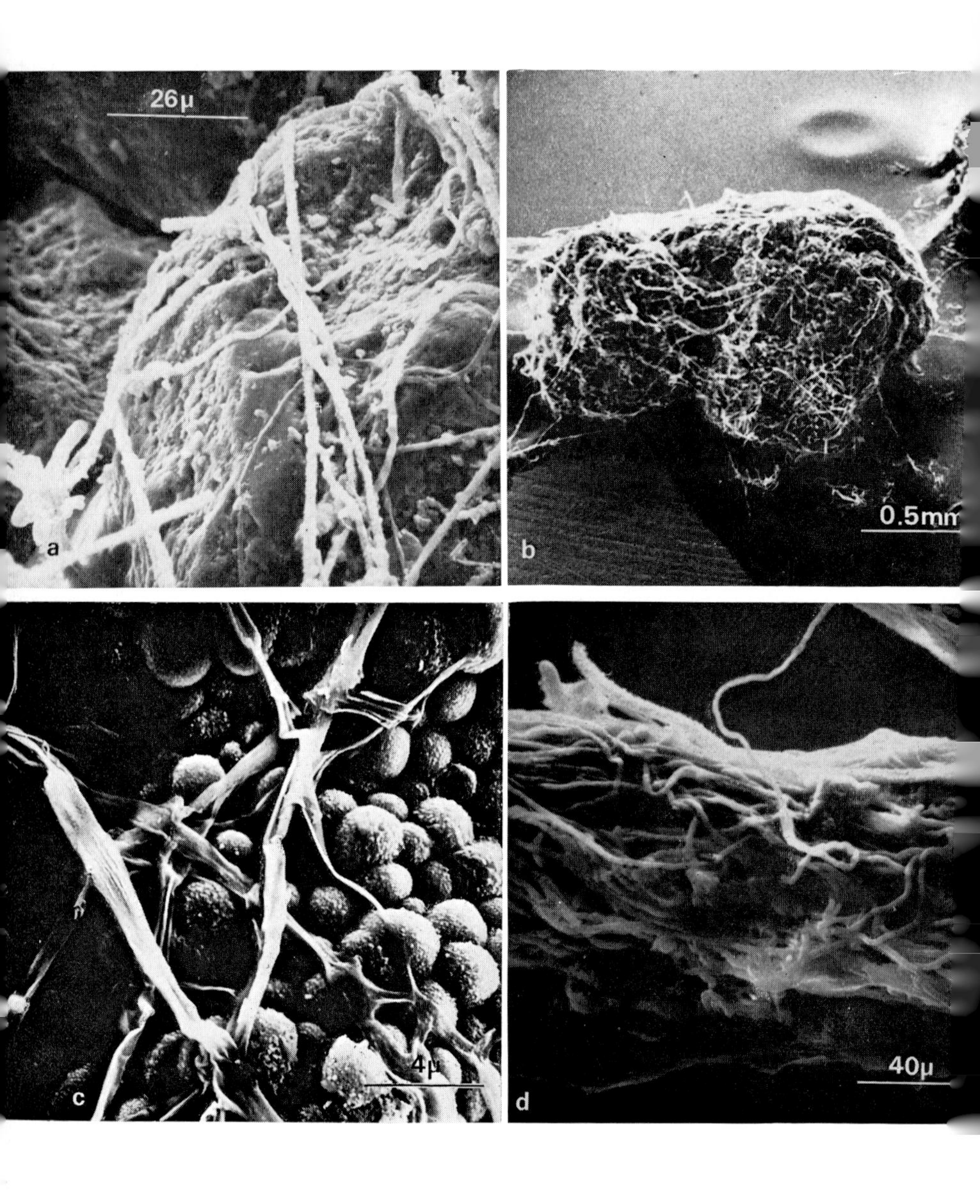

Plate II. *Structures typical of soil fungi as seen with the scanning electron microscope.* (a) *Diffuse, spreading hyphae in soil,* (b) *sclerotium of a* Sclerotinia *sp.,* (c) *surface cells of a* Sclerotinia *sclerotium,* (d) *a rhizomorph removed from the humus layer of a pine forest soil*

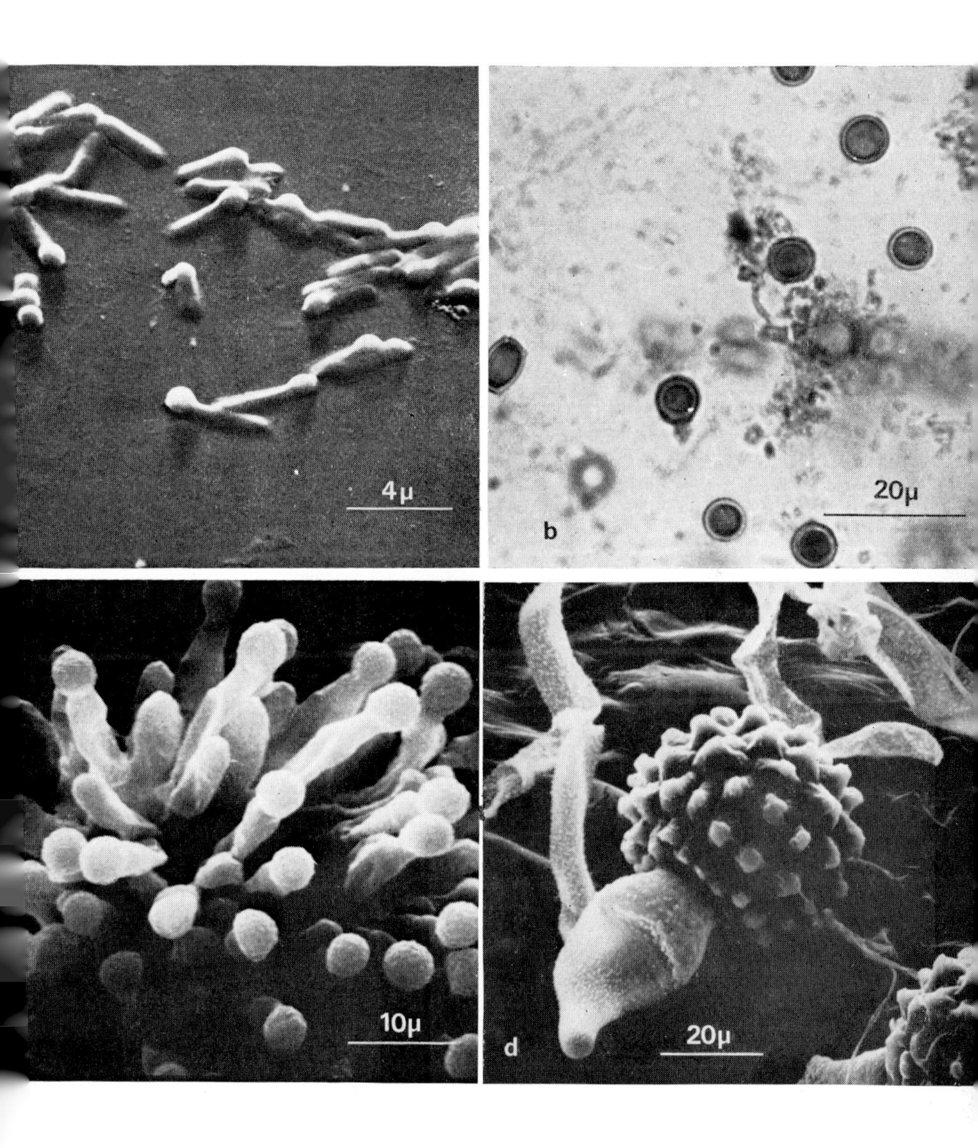

Plate III. *Spores associated with survival or dispersal of soil micro-organisms.* (a) *Spores of* Bacillus sphaericus *still enclosed in the vegetative cells (scanning electron micrograph),* (b) *chlamydospores of* Fusarium oxysporum *(light micrograph),* (c) *conidia of* Aspergillus *(scanning electron micrograph),* (d) *zygospore of* Zygorhynchus moelleri *(scanning electron micrograph)*

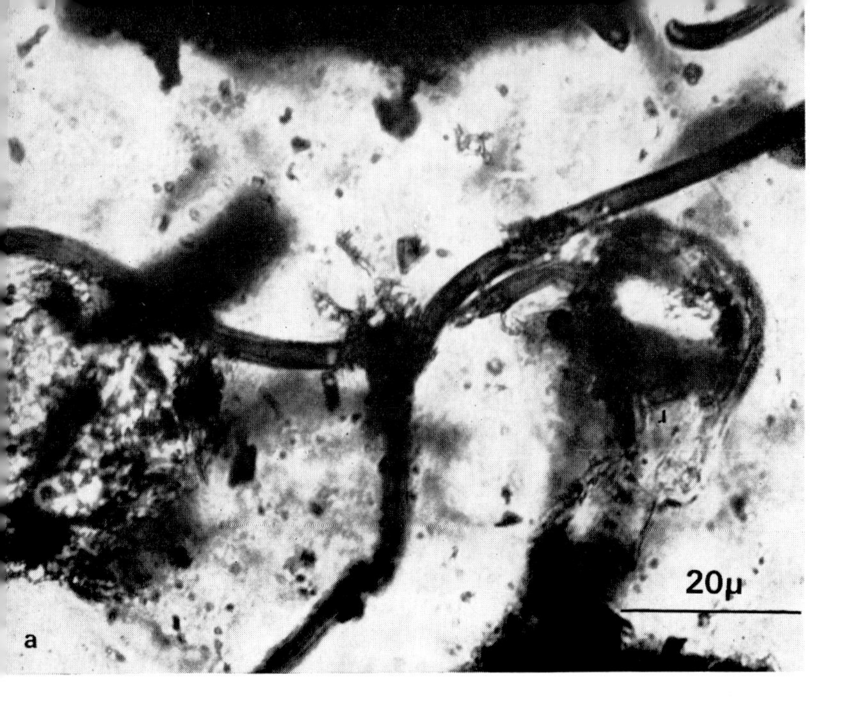

Plate IV. *Light micrographs of thin sections of a resin impregnated forest soil (podzol).* (a) *Humus particles colonised with dark pigmented fungal hyphae,* (b) *mite faecal pellets inside a decaying pine leaf from the humus layer,* (c) (*opposite*) *mineral particles from the* B_2 *horizon showing deposition of sesquioxides on the particle surfaces.* (d) *Electron micrograph of an ultra-thin soil section mucigel containing bacteria (B) sur-*

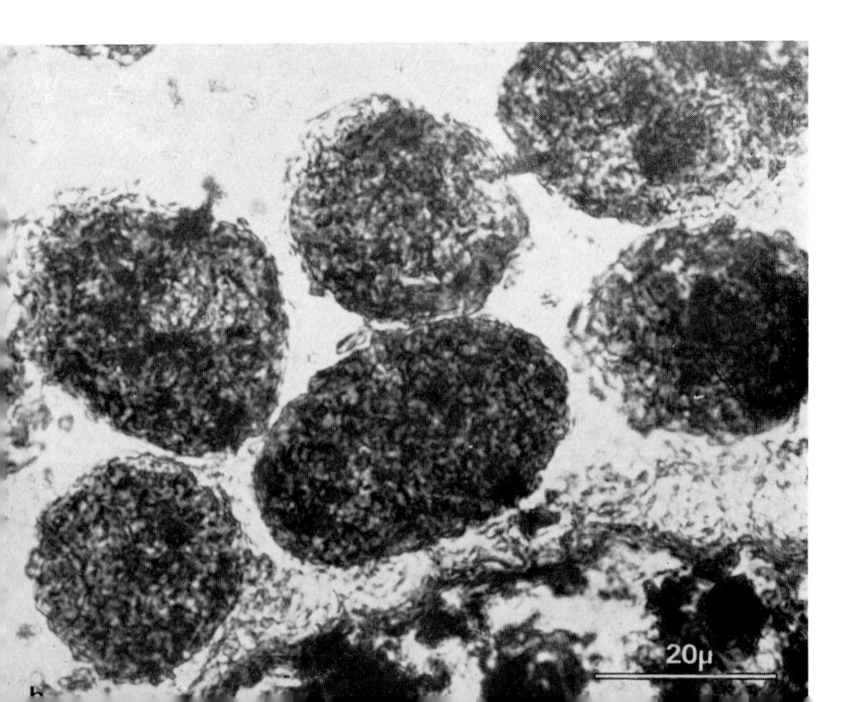

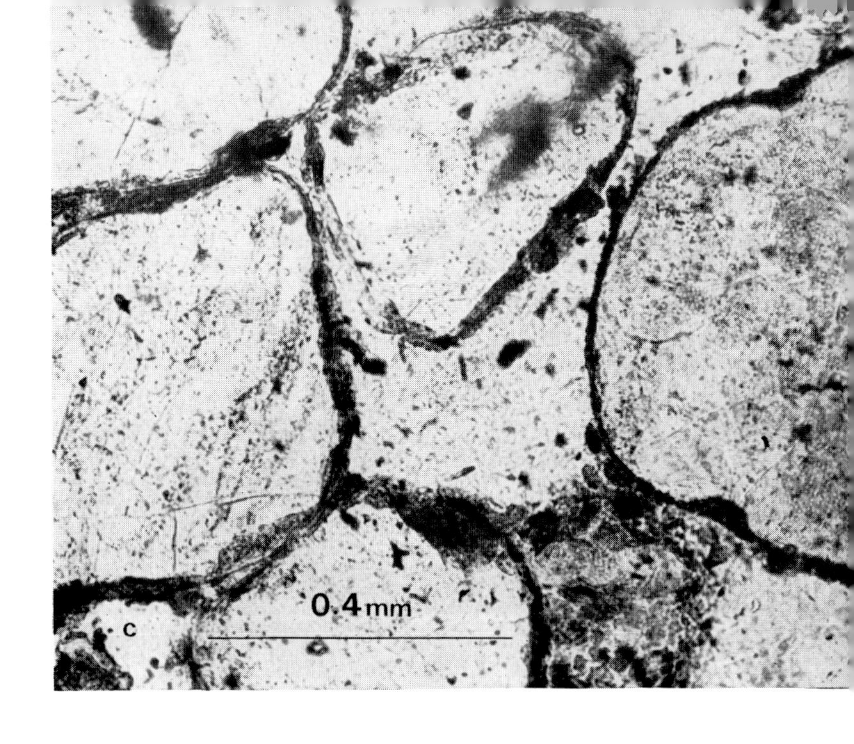

rounding barley roots. Note the dark iron particles (I) which are separated from the root surface (R) by the mucigel. (Jenny and Grossenbacher[233])

(*Plates IV* (a), (b) *and* (c) *are photographs of soil sections prepared by D. P. Nicholas. Plate IV* (d) *is reproduced by permission of H. Jenny and the Soil Science Society of America*)

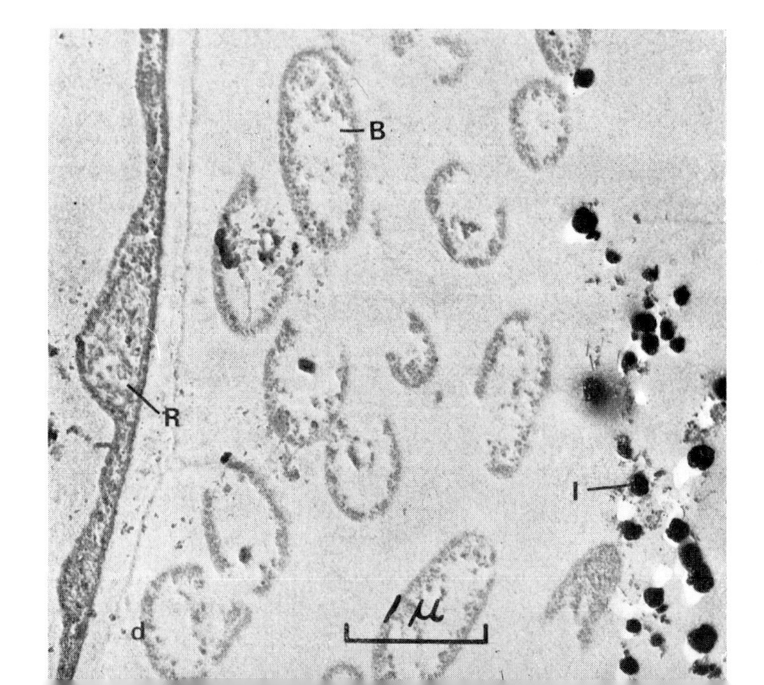

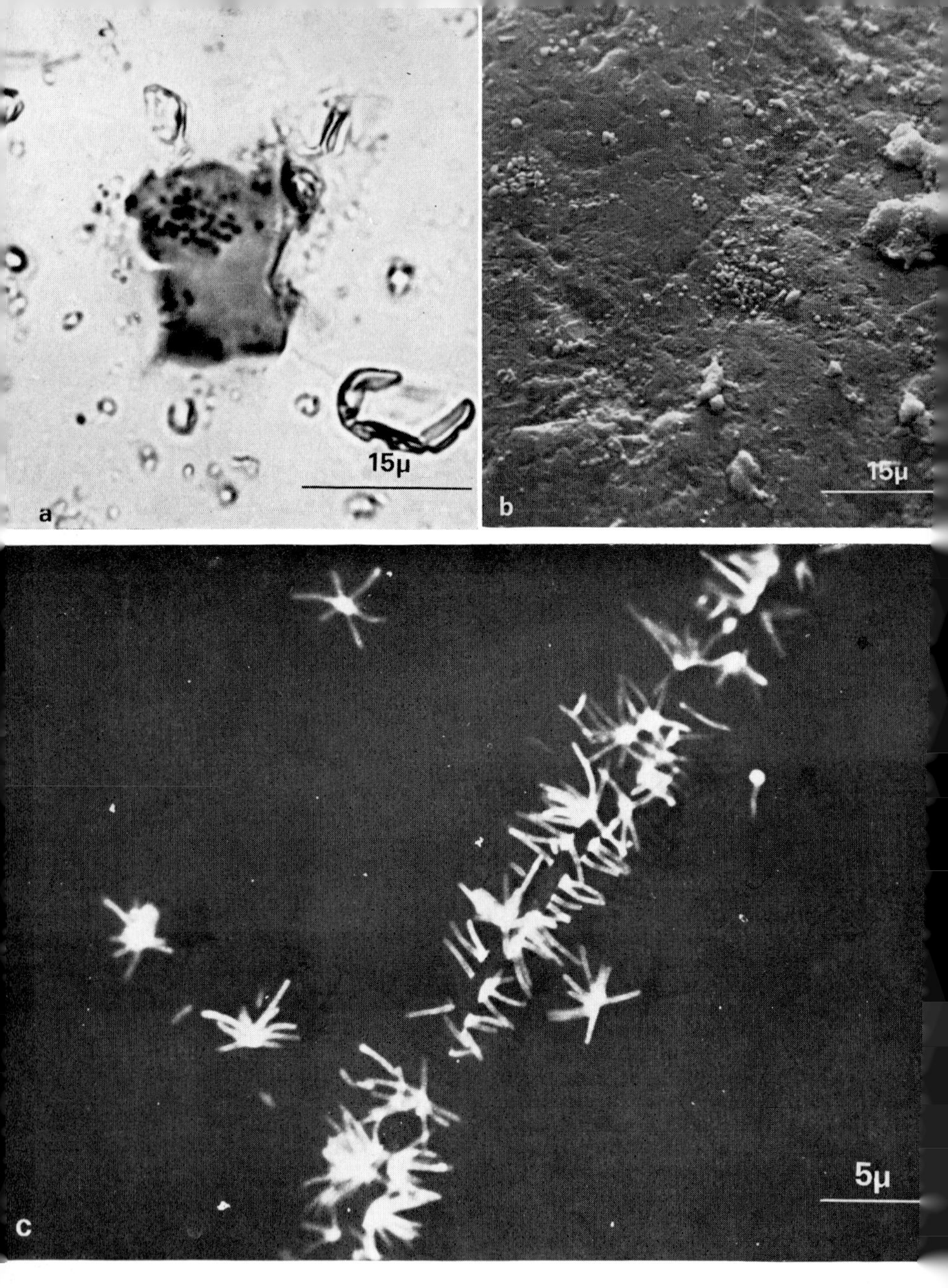

Plate V. *Bacteria in soil detected by direct observation methods.* (a) *Bacteria on a sand grain stained with phenol aniline blue as seen with the light microscope,* (b) *a similar bacterial colony seen with the scanning electron microscope* (*Gray*[154]), (c) *cells of* Rhizobium japonicum *stained with a specific fluorescent antiserum associated with unstained hyphae of* Penicillium janthinellum. *Note the apparent polar attachment of the cells, seen also in Plate VIII* (c)

(*Plate V* (b) *is reproduced by permission of the American Association for the Advancement of Science*)

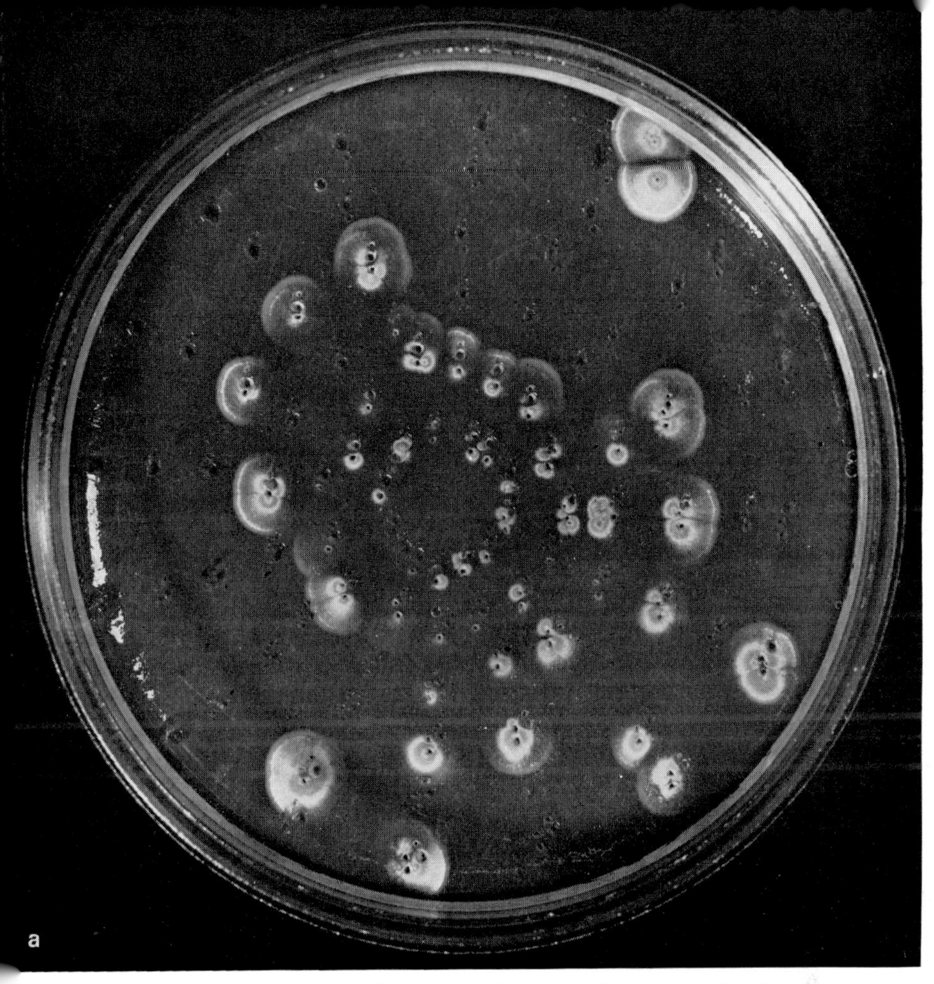

Plate VI. *Growth rate of soil micro-organisms.* (a) *An example of a Stotzky replica plate showing the outgrowth of a* Streptomyces *sp.* (*see p.* 63 *for explanation*), (b) *cells of* Bacillus subtilis *labelled with the fluorescent brightener Calcofluor white. This brightener is transmitted to daughter cells in visible amounts so that the rate of development of labelled bacterial colonies can be followed*

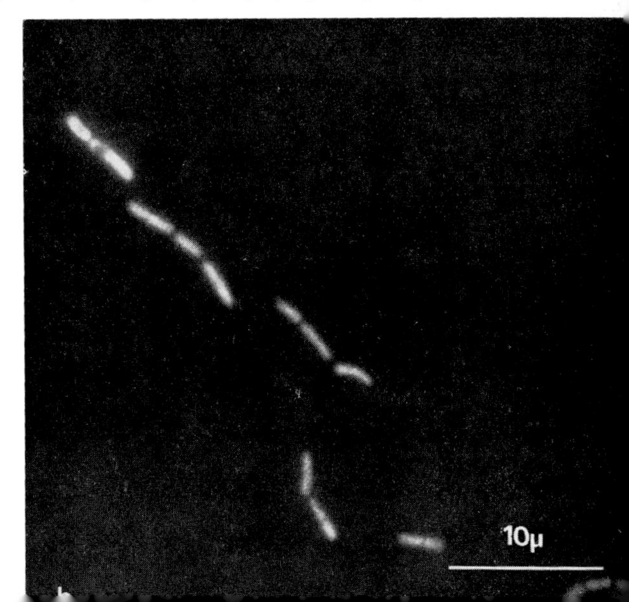

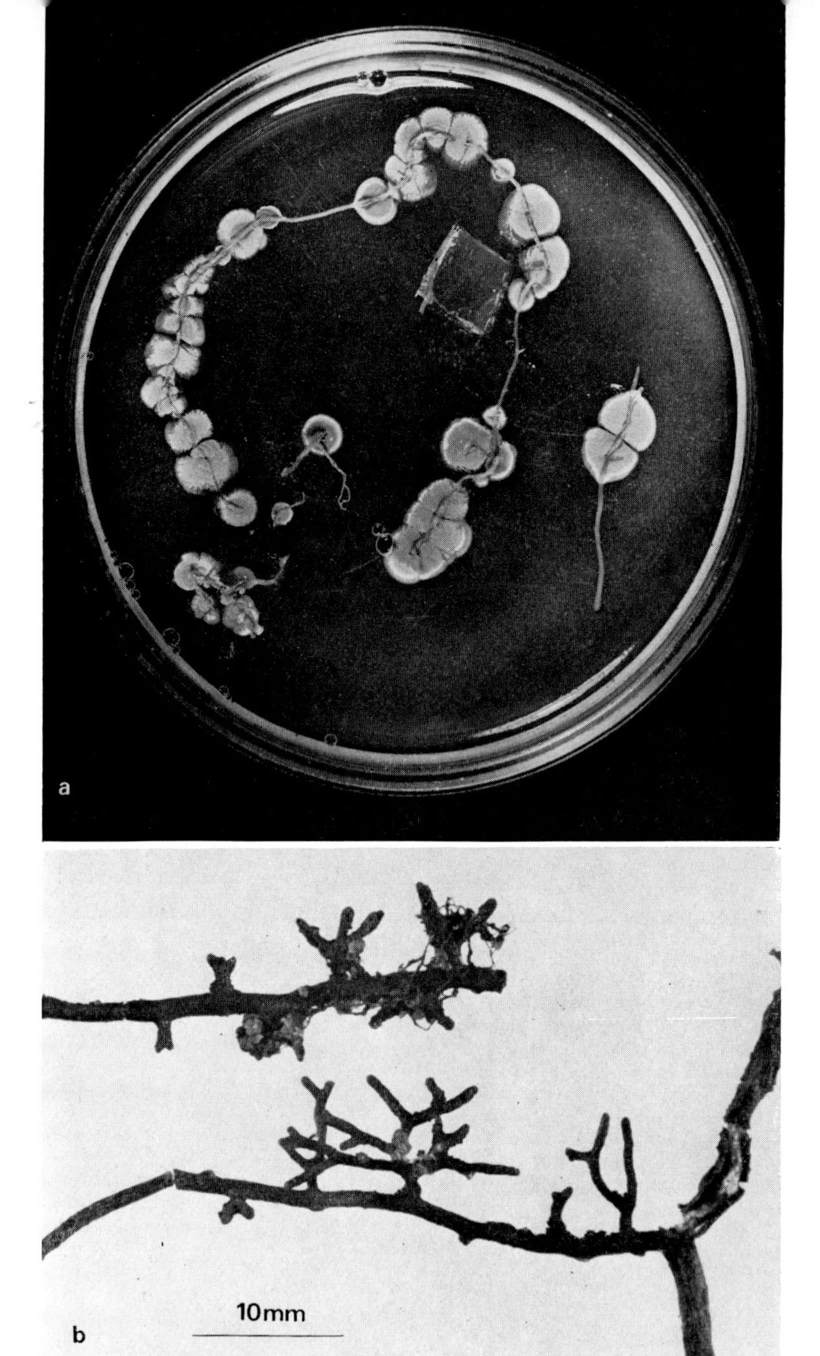

Plate VII. *Inter-relationships of soil micro-organisms with plant roots.* (a) *The development of saprophytic* Streptomyces *sp. from the root surface of* Ammophila arenaria *after placing the roots on a selective isolation medium,* (b) *mycorrhizal rootlets of* Pinus nigra laricio, (c) (*opposite*) *nodules on the roots of* Vicia faba *formed by* Rhizobium leguminosarum, (d) *nodules on the roots of* Alnus glutinosa *formed by actinomycete-like organisms*

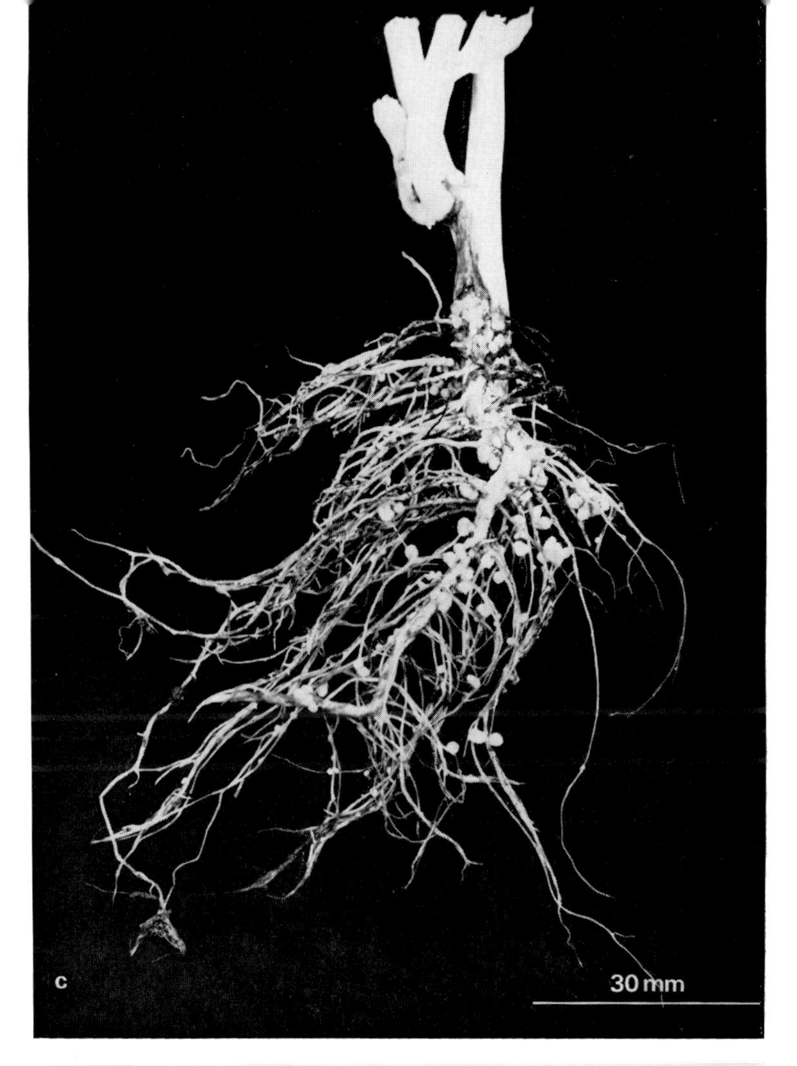
c
30 mm

45mm
d

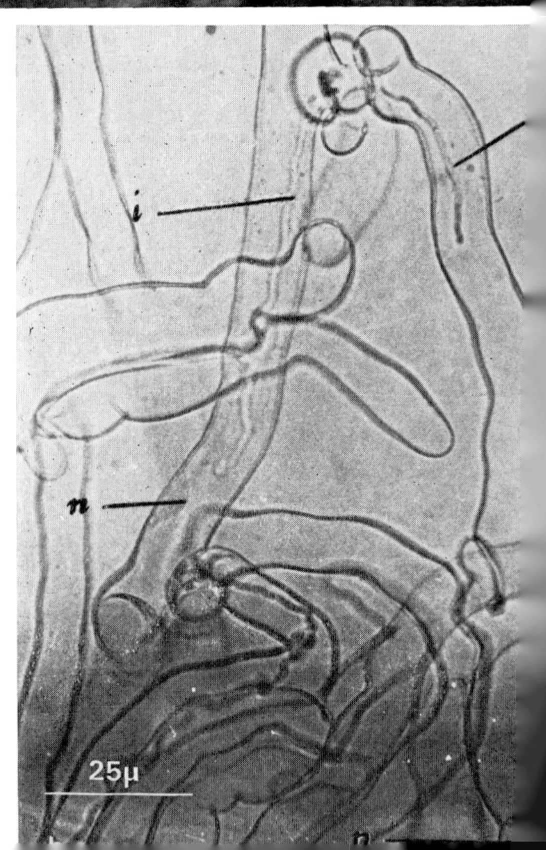

Plate VIII. *Early stages in the formation of legume root nodules.* (a) *root hairs of clover curled by* Rhizobium trifolii, (b) *the formation of an infection thread* (*i*) *in root hairs of* Trifolium fragiferum (*n*) *is the root hair nucleus* (*Nutman*[337 338]), (c) (*opposite*) *electron micrograph of a longitudinal section through an infected root hair of* Trifolium parviflorum *showing the beginning of an infection thread in the fold of the curled hair tip. Some parts of the thread are also seen nearer the base of the hair. Bacteria may be seen in the infection thread and attached to the outside of the hair*[429] (*Sahlman and Fåhraeus*)

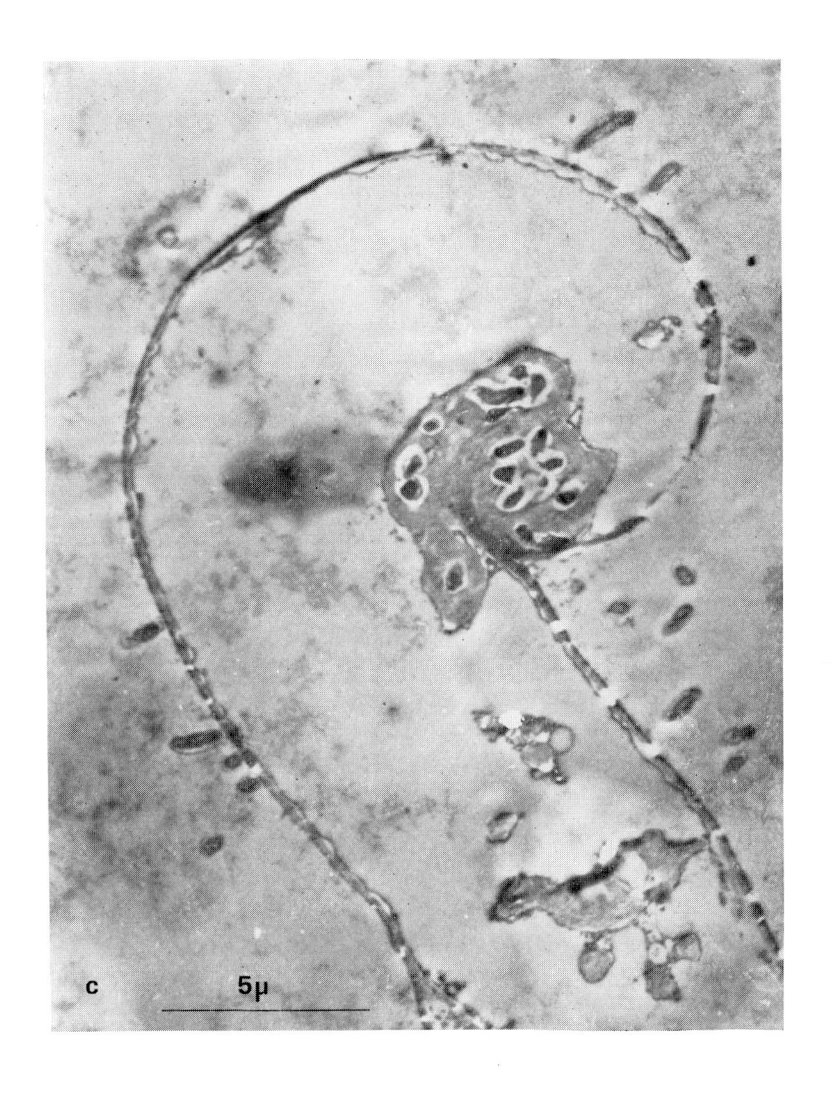

(*Plate VIII* (a) *is reproduced by permission of P. S. Nutman and the Regents of the University of California, VIII* (b) *by permission of P. S. Nutman and Butterworths Scientific Publications and VIII* (c) *by permission of G. Fåhraeus and the Society for General Microbiology*)

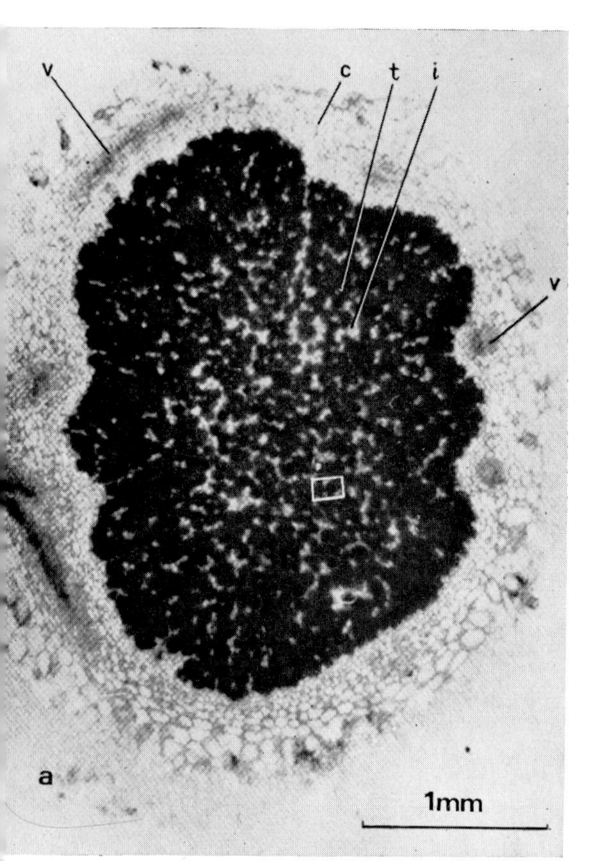

Plate IX. *Internal structure of root nodules of legumes and non-legumes.* (a) *Stained section of a soybean nodule showing cortex* (*c*), *vascular elements* (*v*), *central tissue cells* (*t*), *empty interstitial cells for ventilation* (*i*). *Inset frame is the approximate location of Plate* IX (b). (b) *Electron micrograph of a thin section of a portion of a central tissue cell showing bacteroids* (*b*) *within a membrane envelope* (*e*) *lying in the featureless host cytoplasm* (*c*). *Mitochondria of the host* (*m*) *are concentrated adjacent to the interstitial cell* (*i*). (c) (*opposite*) *Transverse section through a young root-nodule of* Alnus glutinosa *stained with Heidenhain's iron haematoxylin. In some nuclei, mitosis*

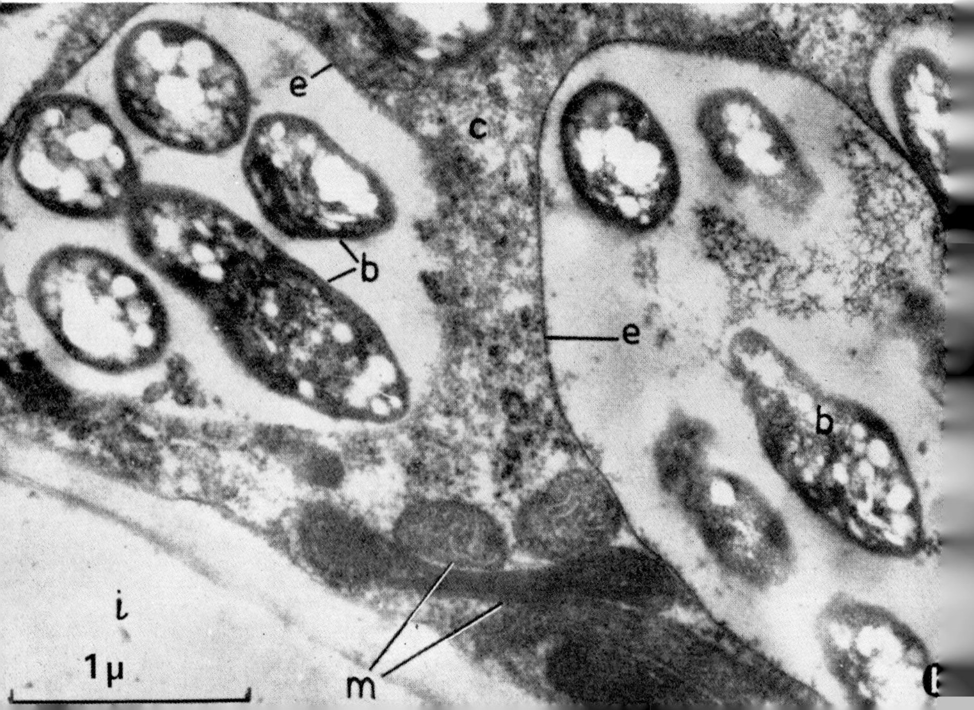

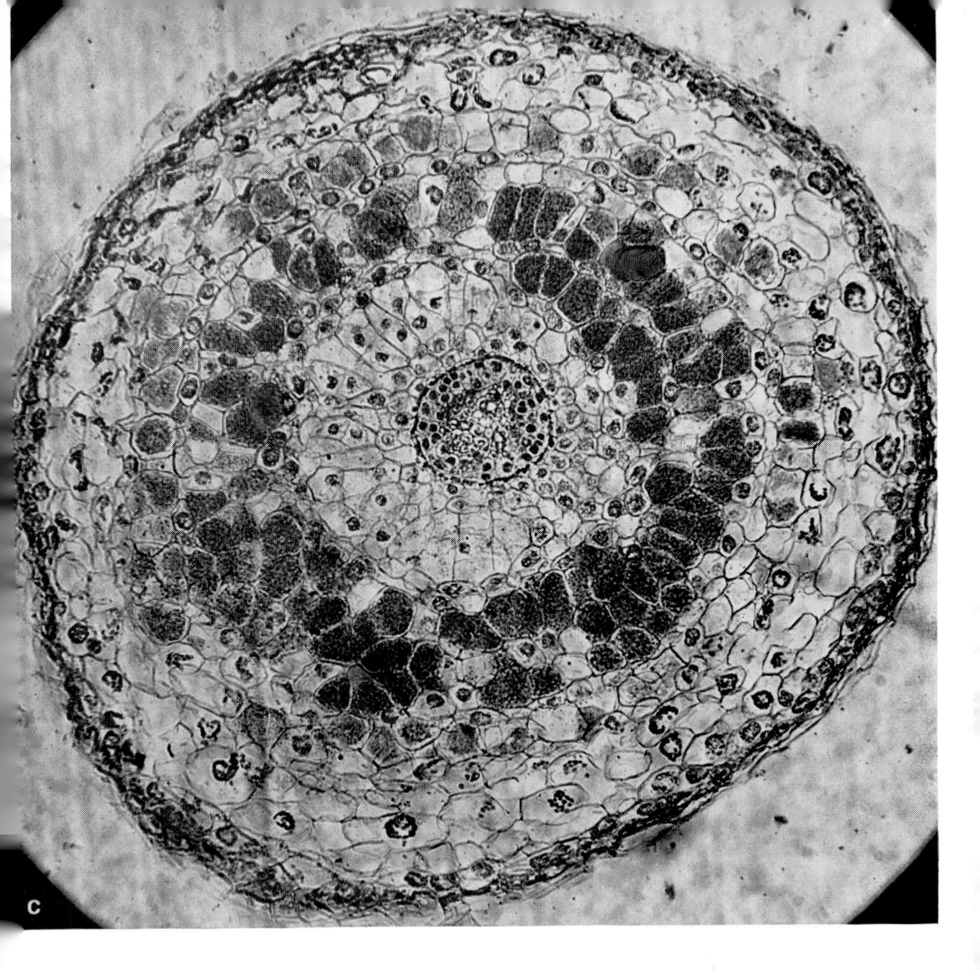

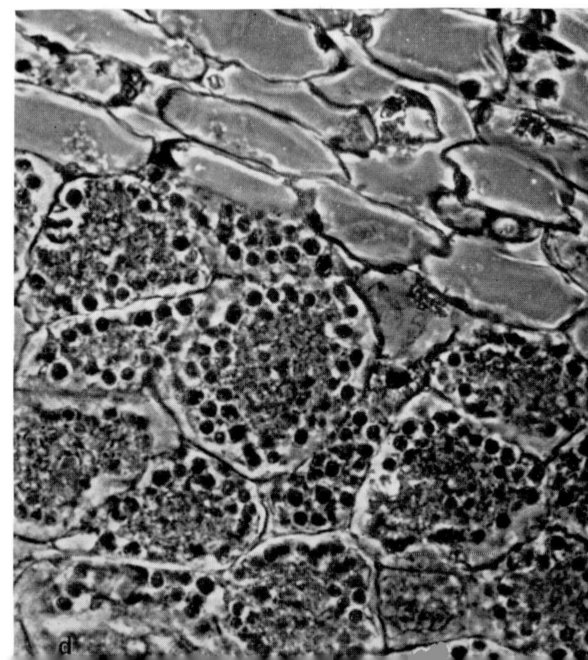

and separate chromosomes are visible. Infected cells are present in a distinct zone about halfway through the cortex. (d) *Cortical parenchyma cells of* Alnus glutinosa (*phase contrast*) *with spherical vesicles. The central part of the host cell is occupied by hyphae; the vesicles are situated peripherally in the cytoplasm. The enlarged cells containing the endophyte are probably tetraploid*

(*Plates IX* (a) *and IX* (b) *Bergersen,*[29] *are reproduced by permission of F. J. Bergersen, D. J. Goodchild and the International Society of Soil Science, and Plates IX* (c) *and IX* (d) *Becking,*[27] *by permission of J. H. Becking and Centraal Stikstof Verkoopkantoor*)

Plate X. *Effects of micro-organisms on plant growth and soil structure.* (a) *Six-month-old* Alnus glutinosa *plants grown in nitrogen-poor, sterilized soil and lacking nodules compared with* (b) *plants of the same age grown in non-sterile soil and possessing nodules* (*Becking*[27]), (c) (*opposite*) *aggregates of sand particles from a dune soil, formed by the binding of grains by fungal hyphae*

(*Plates X* (a) *and X* (b) *are reproduced by permission of J. H. Becking and the Central Nitrogen Sales Organization Ltd.*)

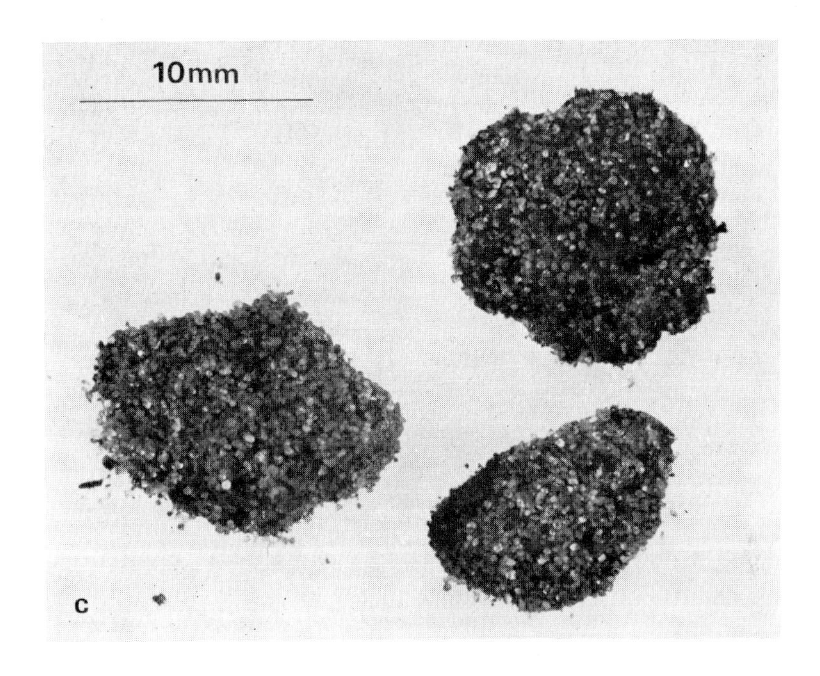

Plate XI (*below*). *Interactions between soil organisms.* (a) *A host cell of* Escherichia coli *with three* Bdellovibrio *parasites attached to it and one free*

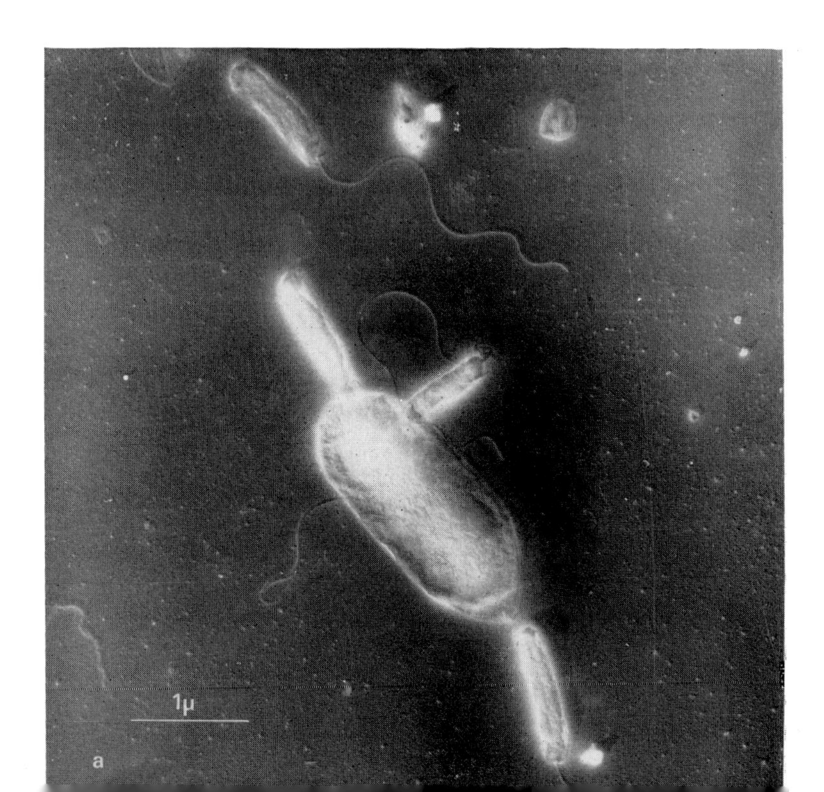

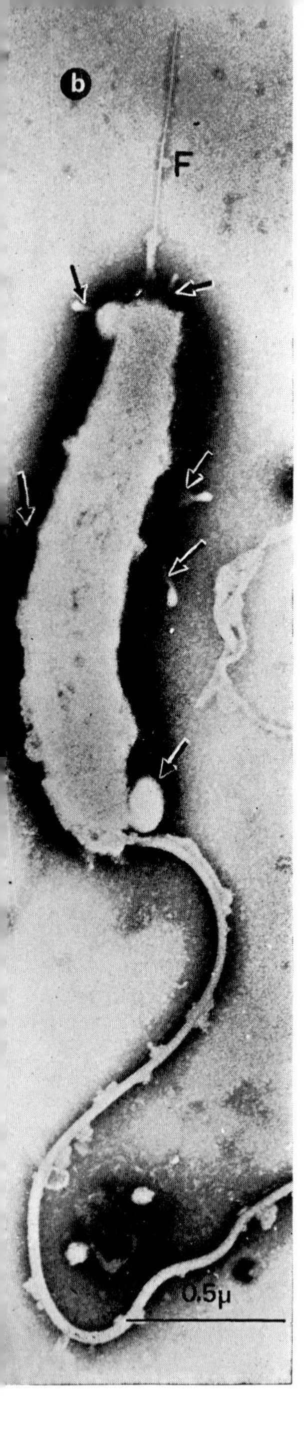

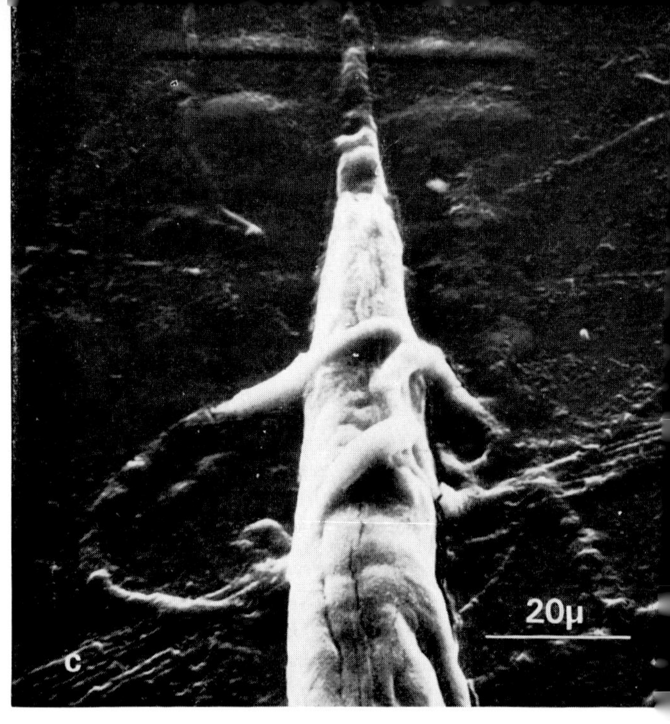

Plate XI (cont.). (b) *a curved rod-shaped cell of* Bdellovibrio bacteriovorus. *Note the straight filament (F) at the attachment pole, the single flagellum at the opposite pole and various appendages (arrowed) (Shilo*[445]), (c) *a scanning electron micrograph of the tip of a nematode caught by the sticky hyphal network of the predacious fungus* Arthrobotrys, (d) *a scanning electronmicrograph of the three-dimensional, sticky network of* Arthrobotrys

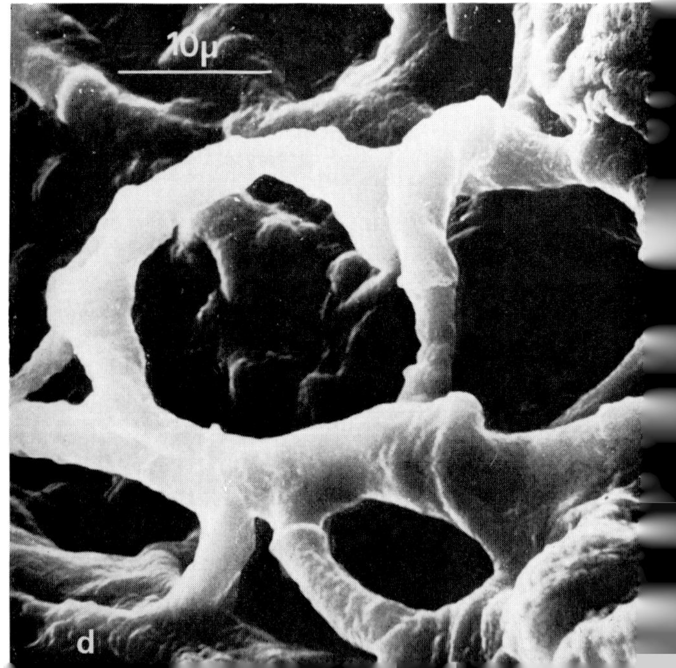

that the present groupings may need revision after further investigations. At present there are six main cross-inoculation groups that are widely recognised and six species of *Rhizobium* which infect them[234]. These are the clover group which is infected by *R. trifolii*, the bean group infected by *R. phaseoli*, the pea-vetch group infected by *R. leguminosarum*, the lupin group infected by *R. lupini*, the soyabean-cow pea group infected by *R. japonicum* and the medic-sweet clover group infected by *R. meliloti*.

(*i*) *Rhizobia in the Soil* When the nodules degenerate, bacteria are liberated into the soil. In the absence of a suitable host, they gradually die out, although Vincent[529] found that they could survive for at least 10 years. It is not surprising, therefore, that soils in which legumes have never grown are deficient in rhizobia. If host plants are present, the rhizobia are stimulated and large populations occur in the rhizosphere (up to 10^5 to 10^7 cells per gram of soil); they also occur in the rhizosphere of non-legumes but not in such large numbers. A specific attraction factor from the host root has been searched for but so far no substance peculiar to legumes has been found. Substances such as biotin are secreted in large quantities by legume roots, but their stimulatory action on the soil microflora is unspecific. Once nodulation has occurred, further nodule formation is inhibited, possibly by the secretion of gibberellin-like substances from the root. Root secretions may also account for the differences between plants which nodulate quickly and those which do so more slowly. These responses are controlled by genetic factors, which may act by altering the rate of build-up of infection-promoting substances.

Other organisms in the soil can influence rhizobia, and Holland and Parker[215] suggested that the failure of nodulation in newly cleared soils of Western Australia was caused by the inhibition of rhizobia by antibiotics from fungi growing on the plentiful organic debris in these soils.

(*ii*) *Development and Structure of the Nodules* Infection of legume roots normally takes place through the root hairs which become elongated, deformed and curled. The elongation of the hairs may be induced by indole acetic acid which rhizobia can synthesise from a precursor, tryptophan, secreted by the root, but the cause of the hair deformation is unknown. The infection usually occurs at the growing point of the hair and it has been suggested that the bacteria attach themselves there by means of fimbriae[105].

Build-up of bacterial metabolites, perhaps in the mucigel or in the folds of the curled hair (Plate VIII (a)) then induces further changes. Fåhraeus and Ljunggren[130] have suggested that the bacterium and plant together synthesise the enzyme polygalacturonase, which weakens the mainly pectic wall of the root-hair growing-point. They also suggested that polygalacturonase formation is induced by polysaccharides produced by the bacterium; as different polysaccharides are characteristic of different rhizobia, this could provide a basis for the specificity of the interrelationship. However, Lillich and Elkan[269] found that levels of polygalacturonase activity in extracts from roots inoculated with rhizobia were not significantly greater than those in uninoculated control plants. Some of the suggested interactions between the bacteria and the host root prior to infection are shown in Fig. 21.

The weakening of the cell wall results in the formation of an infection thread (Plate VIII (b)), but the mechanism of its production is not clear. Dart and Mercer[105] supposed that the fibril meshwork of the root-hair wall was sufficiently loose to allow the passage of rhizobia. This hypothesis would require that the infection thread was laid down by the root hair as a defence against the rhizobia, rather than being a normal continuation of hair growth. Nutman[336], however, suggested that the direction of root-hair growth was reversed, resulting in an invagination of the wall. Certainly an infection thread can be seen inside the root hair, connected to the cell wall and it can be shown that it is made from plant cell wall components (Plate VIII (c)).

The infection thread grows through the root hair, generally towards the nucleus which shows increased activity at this time, until it reaches the cell boundary. Here it induces the formation of a new thread which passes into the next cell and this continues until it reaches the root cortex. In the cortex, cells with twice the normal complement of chromosomes (tetraploid), normally associated with the sites of lateral root formation, are infected. The thread bursts and the rhizobia are liberated into the cytoplasm of a cell. This and neighbouring cells then divide repeatedly until a mature nodule is formed, composed of internal infected tetraploid cells and external uninfected diploid cells. The infected cells are swollen and contain the pigment haemoglobin, while the uninfected ones are differentiated into cortical, meristematic and vascular tissue (Plate IX (a)).

The bacteria, once liberated from the infection thread, multiply rapidly and become swollen and irregular in shape, forming 'bacteroids'. The bacteroids are irregularly shaped cells, probably lacking most, if not all, the normal cell wall components, and bounded only by a cytoplasmic membrane. They are enclosed singly or in packets of two to four in folds of the host cell's membrane system (Plate IX (b))[30, 103]. It is thought that nitrogen is fixed at this stage, possibly at sites where

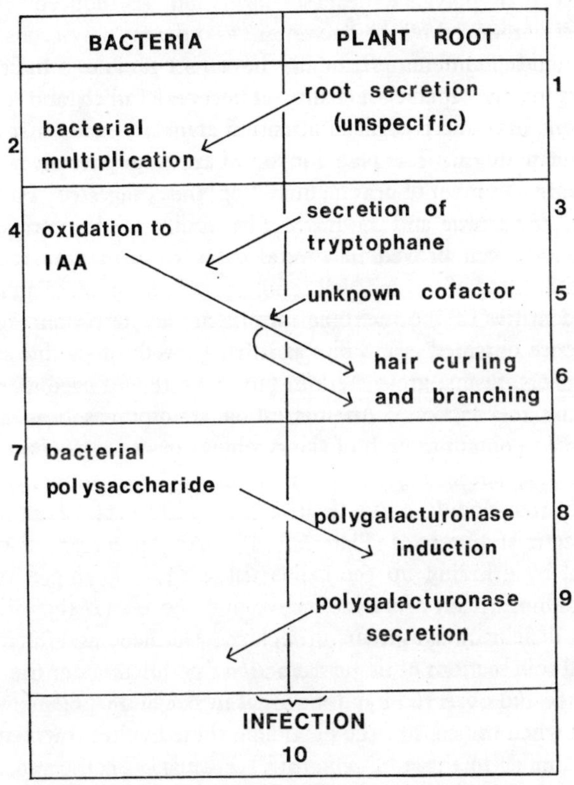

Fig. 21. *Possible interactions between* Rhizobium *and legume roots before infection* (*Nutman* [338])
(*Reproduced by permission of P. S. Nutman and the Regents of the University of California*)

the membranes of the bacteroid and plant cell are in close proximity. The bacteroids of effective strains which can fix nitrogen differ from those of ineffective strains which are unable to fix nitrogen. In the former, the nuclear area elongates, divides and becomes surrounded by a clear perinuclear area; outside this is a more granular cytoplasm containing 'particles' associated with oxidation/reduction reactions. In the ineffective strains, there is no clear perinuclear area and there are large accumulations of glycogen, indicative of both excess carbohydrate and nitrogen deficiency. The bacteroids die as the root becomes senescent and only the bacteria which remain in the infection thread survive to be liberated into the soil.

(b) Bacterial Associations with Non-legumes

Certain non-leguminous plants also have root nodules which contain micro-organisms capable of fixing atmospheric nitrogen[36]. These associations have received little attention compared with that given to the legume nodules. The plants involved are mainly woody perennials and occur in a number of genera, including *Alnus*, *Casuarina*, *Ceanothus*, *Eleagnus*, *Hippophae* and *Myrica*. The ability of the nodules to fix nitrogen has been proved in several cases by using the radioactive tracer ^{15}N.

The identities of the microbial symbionts are uncertain. Most of the evidence obtained is indirect as it has proved impossible to isolate the micro-organisms, grow them in pure culture and produce effective nodules by re-infection. An interesting attempt has been made by Becking[27] to obtain growth of the symbiont in tissue cultures of alder nodules. He succeeded in obtaining intracellular actinomycete-like growth in root-nodule callus tissue but failed to observe formation of the characteristic vesicles (Plate IX (d)). Although nodules could be produced by grinding up the callus tissue and applying it to sterile *Alnus* seedlings, they were ineffective and the plants showed typical nitrogen deficiency symptoms. Becking, De Boer and Houwink[28] examined thin sections of tissue from *Alnus* nodules under the electron microscope and observed hyphae similar in size and cytology to actinomycetes; when the cells of the host died, these hyphae fragmented in a manner similar to those of *Nocardia*. Furman[137] observed *Streptomyces*-like organisms in *Ceanothus* nodules and Wollum, Youngberg and Gilmour[573] isolated a *Streptomyces* sp. from *Ceanothus* which caused distortion of root hairs on re-inoculation of aseptically grown plants.

Infection of the root takes place through the root hairs, and deformation of the root hairs of *Alnus* and formation of infection threads like those seen in legume associations have recently been reported by Becking[27]. He also showed that the nodule was a modified lateral root originating from protoxylem and protophloem strands within the pericycle of the parent root, and containing an endodermis distinct from that of the parent root. The modified roots are dichotomously branched so that the mature nodule cluster is coralloid in appearance (Plate VII (d)). Clusters the size of a clenched fist may form, sometimes coloured orange with an anthocyanin pigment. There are no root hairs or root caps on the nodule branches but there is usually a corky layer formed in the surface layers of the nodule. The endophyte is not uniformly distributed in the nodule but occurs in a ring or series of concentric rings of the mid-cortex parenchyma cells. The infected cells are larger than the uninfected ones, suggesting that they may have become polyploid as a result of infection[27] (Plate VII (c)).

The exact site of nitrogen fixation is unknown but the membranes of the plant cell surround the invading micro-organism and might be the site of fixation. The micro-organisms probably obtain supplies of organic carbon from the root, but how dependent they are upon this is not known, for there is no information on their activity outside the root.

(c) *Fungal Associations with Plant Roots*

Fungi form symbiotic associations with many plants, ranging from the minute gametophytes of primitive vascular plants to trees such as beech. All associations which involve a prolonged period of healthy interaction between fungi and roots are known as mycorrhiza. A number of general types can be recognised on the basis of their morphological and physiological characteristics, and in this chapter the benefits derived by the fungal partners will be considered.

(*i*) *Ectotrophic Mycorrhiza* In ectotrophic mycorrhiza, the bulk of the fungal material is found outside the host root. These associations are formed mainly with trees, especially those in the families *Pinaceae*, *Betulaceae*, and *Fagaceae*. The roots infected are the short laterals and the infection results in their increased branching; in pine the infected roots branch dichotomously and in beech, repeated branchings occur at right angles to each other. A large amount of the fungus material forms a sheath (20-40 μm thick) surrounding the root, while internal penetration of the host is intercellular and confined to the outer cortex cells which become rather elongated (Fig. 22a).

Infection occurs when seedlings are a few weeks old and seems to coincide with the onset of photosynthesis, when the first true leaves unfold. There is evidence that high light intensity and high carbohydrate content in the host cells promote the infection. The inoculum for infection can originate from other infected roots or spores of the fungus in soil. The fungi involved are mainly basidiomycetes, such as

Amanita, Boletus, and *Tricholoma.* A few fungi in other groups can form ectotrophic mycorrhiza, the best known being *Cenococcum graniforme,* an imperfect fungus which has a very wide host range. Generally, the fungi are not completely host specific and there is no evidence that root exudates have any specific effect on potential mycorrhizal fungi in soil. It is no easy task to prove that a fungus is capable of forming a mycorrhiza. Seedlings of a suitable plant must be grown in aseptic conditions (pine seeds are often used because they can be sterilised more easily than those of angiosperms) and when they are inoculated with the fungus, a typical mycorrhiza must form. However, in such artificial conditions a negative result does not necessarily indicate that the infection will not occur in the field.

Physiological studies of the fungi involved, especially those studies undertaken by Melin and his co-workers in Sweden[39, 320], have shown that most of the fungi need a supply of simple carbon compounds and are incapable of breaking down more complex substances like cellulose. This is in marked contrast to most other basidiomycetes which are noted for their ability to break down complex materials. Many need vitamins and are inhibited by phenolic substances present in extracts of plant litter. Thus they seem singularly ill-equipped for a successful saprophytic existence in soil or litter. One might conclude, therefore, that these fungi escape from competition for substrates in soil and obtain instead a supply of nutrients from the root. Using ^{14}C-labelled sucrose and other sugars, Lewis and Harley[265-267] were able to show that transfer from the host to the fungus occurred, the labelled material eventually appearing inside the hyphae, often in storage products such as mannitol. These products were not significantly utilised by the root and therefore were of clear value to the fungus.

(ii) Endotrophic Mycorrhiza of Orchids In the endotrophic mycorrhiza of the *Orchidaceae,* most of the fungus is within the root tissues, with a few connecting hyphae running into the soil. Some of the hyphae grow inside the cells in the outer cortex (Fig. 22b). A number of unrelated fungi can form associations with orchid roots, among them *Rhizoctonia solani* (a sterile fungus), several basidiomycetes and Fungi Imperfecti.

The relationships between the root and the fungus are not completely understood. It is known that the hyphae growing inside the host cells are occasionally digested and there is evidence that the host produces substances toxic to the fungus, one such substance being a phenolic compound known as orchinol[144]. Both of these observations suggest that the relationship is an unstable one. This is also supported by the fact that some of the fungi forming mycorrhiza with orchids, e.g.

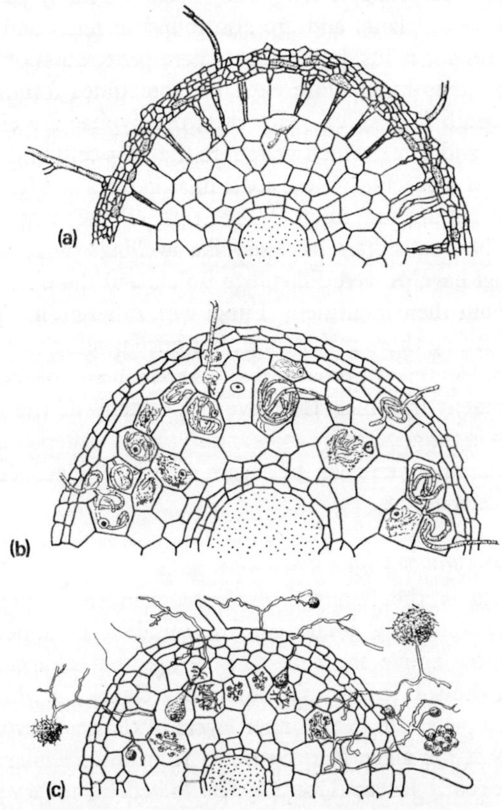

Fig. 22. (a) *Transverse-section of ectotrophic mycorrhizal rootlet showing fungal sheath and intercellular penetration.* (b) *Transverse section of orchid mycorrhiza showing fungal penetration and intracellular digestion.* (c) *Transverse section of vesicular-arbuscular mycorrhiza showing penetrating hyphae, arbuscular haustoria, digestion and vesicles. Fruit bodies are shown externally on the extramatrical mycelium* (*Harley*[186])

(*Reproduced by permission of J. L. Harley and the Regents of the University of California*)

Rhizoctonia solani and *Armillaria mellea*, are pathogenic to other plant roots. It seems likely that the fungus benefits by the limited utilisation of nutrients in the host cells.

(iii) Endotrophic Mycorrhiza of the Vesicular-arbuscular Type
Although comparatively little is known about these mycorrhiza, they are the most widespread of all. They occur on many perennial and annual herbaceous plants and are also found in ferns and liverworts. Most of the fungus is inside the root, where penetration of the cortical cells occurs; some hyphae are external, sometimes forming a rather ill-defined sheath (Fig. 22c). However, the hyphae are characteristically aseptate and those which penetrate the host cells end in repeated branches called arbuscules; these are similar in form and function to the haustoria of pathogenic fungi. The tips of the hyphae growing between the host cells often have sac-like swellings called vesicles.

These fungi have proved difficult to isolate and there has been some confusion about their identities. Fungi with different morphology are capable of forming these mycorrhiza, but once inside the host they are indistinguishable; this is probably due to the influence of the specialised host environment in which they live[195]. Many of the associations involve fungi in the genera *Endogone* and *Pythium*. Interactions between the partners are not yet clear, but again occasional digestion of hyphae by the host cells occurs.

(d) Algal Associations with Plant Roots

Certain plants in the family *Cycadaceae* form roots which are repeatedly dichotomously branched, are coralloid in appearance and sometimes grow above the soil. These may be colonised, while in contact with the soil, by certain blue-green algae, e.g. *Anabaena* and *Nostoc*, which penetrate to the inner layers of the root cortex. Here it is possible that the algae gain nutrients from intercellular slime produced by the root. Infected roots appear to benefit, as they grow larger and live longer than uninfected ones. It has been shown that *Nostoc cycadae* can fix ^{15}N in both the free living state and in association with cycad roots[550].

Pathogenic Relationships between Soil Micro-organisms and Plant Roots

Many soil micro-organisms can invade plant roots and grow in their tissues as parasites. The diseases produced are of great importance in agriculture, forestry and horticulture, as they result in considerable losses and reduction of yields. Diseases which spread through the soil are extremely difficult to control as the micro-organisms are more difficult to detect and eradicate than pathogens attacking the aerial parts of the plant. Also, many pathogens can survive, either actively or passively, for many years in soil in the absence of the host plant.

Of the microbes which parasitise roots, the vast majority are fungi. Some of them are almost completely confined to the host tissues during the active phases of their life, while others can grow as saprophytes in soil for an indefinite period. These marked differences led Garrett[140] to divide soil-borne pathogenic fungi into two groups, the soil inhabitants and the root inhabitants.

Relatively few species of bacteria are root pathogens, those that are being aerobic, non-sporing rods in the genera *Agrobacterium*, *Corynebacterium*, *Erwinia*, *Pseudomonas* and *Xanthomonas*. As with the fungi, there is a considerable range in saprophytic activity and attempts have been made to group them on their behaviour outside the host[56, 99]. Only one actinomycete has been definitely proved to be pathogenic to roots and that is *Streptomyces scabies* which causes scabs on potatoes and other root crops.

(*a*) *Soil-inhabiting Pathogens*

These pathogens are able to survive indefinitely in soil as they have a high competitive saprophytic ability (see Chapter 4), enabling them to compete successfully with other members of the microflora for dead organic matter. If, however, a suitable host is present and conditions promoting infection occur, they will infect roots and also, in some instances, seeds and the lower parts of stems. They may be considered to be primitive parasites, since the host is often quickly killed and specialised relationships do not exist, most of them having a very wide host range. The resistance of a mature host is often sufficient to restrict infection to young or moribund tissues. A good example of a soil-inhabiting, pathogenic fungus is *Pythium mamillatum*, causing 'damping off' of seedlings and which Barton[22, 23] showed to be a highly successful saprophyte, capable of pioneer colonisation of many substrates. Another is *Fusarium culmorum*, which in addition to causing seedling blight of wheat, is also a primary coloniser of wheat straw in soil[427].

The majority of plant pathogenic bacteria do not persist for long in soil in the absence of the host. There are, however, some bacteria which are predominantly saprophytic, their pathogenicity being incidental to their life in soil around the root. Examples are some of the fluorescent pigmented *Pseudomonas* species which cause 'soft rots'. Certain other bacteria, such as *Pseudomonas solanacearum* are widespread in soil and appear to persist in the absence of the host. However, this bacterium has a wide host range and may persist by infecting other plants rather than by saprophytic activity[99]. It also seems likely that *Streptomyces scabies* is capable of a saprophytic existence in soil and should be placed in this group.

(*b*) *Root-inhabiting Pathogens*

These are less able to compete with saprophytic micro-organisms in the soil, but the extent of their saprophytic ability varies, some having none while others can grow outside the root for a limited period. They may survive in soil as resting structures for a considerable time, this being particularly true of fungi forming resistant spores or sclerotia and in such cases, roots exude substances which break the dormancy of these propagules. These microbes are successful parasites because they can grow inside host tissues for a long period of time, causing minimum disturbance and having delayed effects on the host. Garrett[143] considered that there were two main tendencies in the development of specialised parasitism, firstly, an increasing ability to overcome host resistance and secondly, an increasing harmony between pathogen and host. As a consequence, most of these pathogens have a relatively restricted host range.

Two well-known examples of root-inhabiting pathogenic fungi are *Ophiobolus graminis*, which causes 'take-all' of wheat, and *Helminthosporium sativum*, which causes 'foot-rot' of cereals. Some bacterial pathogens have no free life in soil (e.g. *Erwinia amylovora* causing 'fire blight') or else the soil acts only as a vehicle for their transmission from infected seed to developing seedling (e.g. *Xanthomonas malvacearum*, causing cotton blight). Others spend most of their life inside plants but can exist in soil, in a declining saprophytic phase, for a short time.

As with all attempts to categorise living material, there are a few organisms which do not conform and will not fit satisfactorily into either group. Fungal pathogens, causing 'vascular wilts', spread in the host's vascular system; external effects on the plant are delayed until it is mature, when a sudden irreversible wilting occurs. However, the entry of the fungus is through 'young' roots where some destruction of cortical tissues occurs. It is possible that it is the resistance of the maturing plant that eventually confines the fungus to the vascular tissues. When the plant dies, the fungus can again leave the vascular tissue, spore on the senescent roots and return to the soil. Examples of wilt fungi are *Verticillium albo-atrum* and *Fusarium oxysporum* f. *lycopersici* which infect tomatoes.

Development of Root-microbe Associations

Bearing in mind the wide differences in the intimacy and specificity of associations between micro-organisms and roots, we may summarise them as follows:

(*i*) Associations in which the microbes are saprophytic in soil around roots or on root surfaces. These are unspecific, involving many different micro-organisms and healthy root tissues are not invaded to any great extent.

(*ii*) Associations in which the microbes are saprophytic for much of their life but can invade living root tissues in certain circumstances, e.g. the soil-inhabiting pathogens and certain mycorrhizal fungi.

(*iii*) Association in which the microbes are active mainly or only within living roots. These are specific associations, often between one microbial species and one plant species and include the root-inhabiting fungi, some mycorrhizal fungi and *Rhizobium*.

Despite the lack of evidence, it is tempting to postulate the evolutionary pathways by which these associations develop. One can imagine microbes in the rhizosphere gradually evolving a more specialised relationship and eventually invading the living root as a parasite or symbiont. As it develops the ability to live inside the root (by reducing disturbance of the host's metabolism to a minimum), it may come to rely on the supply of nutrients from the host and will spend less time outside the root in competition with the soil microflora, hence its saprophytic ability will tend to decline with the reduction of selective pressures.

7

Effect of Micro-organisms on Plant Growth

An understanding of the effects of micro-organisms on the growth of plants is essential if we are to exploit the possibilities of altering crop productivity and maintaining soil fertility by the introduction of alien micro-organisms into soils and by harnessing the activity of the indigenous microflora. Micro-organisms can bring about changes in plant growth which may be stimulatory or inhibitory. These may be due to: (*a*) use of microbial metabolites as major nutrients; (*b*) the effect of growth regulators produced by micro-organisms; (*c*) the liberation of otherwise unavailable nutrients from soil organic matter and minerals; (*d*) suppression of plant pathogens; (*e*) the production of phytotoxic substances by saprophytes and parasites; (*f*) the production of enzymes; and (*g*) competition of micro-organisms with plants for essential nutrients.

Not surprisingly, these direct interactions are frequently attributed to organisms in the root region of plants. However, changes taking place in root-free soil are equally important since they may influence subsequent root growth into that soil. Indirect promotion and inhibition of plant growth can be caused by changes in soil structure partly brought about by soil micro-organisms, e.g. aggregation of soil particles and crumb formation.

All plant-microbe interactions are difficult to study because the changes in plant growth are similar to, and may be confused with, changes in growth due to solely physical and chemical events. These events may, in turn, affect microbial growth which can influence plant growth and the environment further. Other difficulties arise because most microbe-plant associations, apart from symbiotic and parasitic ones, are rather unspecific. One is faced not with characterising the interaction of two known organisms, but the vague interplay of one organism, the plant, with some or all of the soil microflora, and although any type of microbe may play an insignificant role in plant growth regulation, the cumulative effect of the whole population could be important.

Use of Microbial Metabolites as Major Nutrients

(a) *Nitrogen*

Although almost any substance produced by micro-organisms and released into the soil might serve as a nutrient for plant growth, many microbiologists have concentrated on interactions involving nitrogen. Nitrogen is particularly important to plants for it contributes to the formation of proteins, but heavily cropped soils lose nitrogen readily through leaching and volatilisation and may become nitrogen deficient unless positive steps are taken to avoid this. In undisturbed soils under long-established climax vegetation, a delicate equilibrium between nitrogen gains and losses exists: little readily available nitrogen is present and any that is added to such soils, especially when accompanied by liming or additions of phosphate, is rapidly utilised by the soil microflora. The resulting stimulation of the microflora causes the liberation of further nitrogen from the previously unavailable supply in the soil humus[48].

Nitrogen is abundant in the earth's atmosphere where it constitutes about 80 per cent of the gases present. Unfortunately, atmospheric nitrogen is not metabolised by most plants, but many micro-organisms are able to fix it and convert it to ammonium compounds, which are available to plants. These micro-organisms include the symbiotic root-nodule bacteria of legumes, nodule organisms associated with certain trees and shrubs, free living bacteria such as *Azotobacter*, *Beijerinckia*, *Clostridium*, *Derxia*, *Nocardia*, *Pseudomonas* and *Rhodospirillum*, many blue-green algae (especially those forming heterocysts) and some yeasts. Conclusive proof of nitrogen-fixation has not been obtained with all these organisms, although they are all able to grow on nitrogen-deficient media.

Other nitrogen conversions in soil that provide nutrients for plants include nitrification and the deamination of proteins. These processes have been dealt with elsewhere (pp. 157, 93 and 105) and in this chapter we shall restrict ourselves to a consideration of some of the nitrogen-fixing organisms.

(*i*) *Azotobacter*. *Azotobacter* is the most familiar of the free-living nitrogen-fixing bacteria and has been credited for a long time with a major role in soil. As we shall see, its importance may have been over-emphasised.

Azotobacter cells are variable in shape. Some species have large rod-shaped cells, while others are coccoid or yeast-like in appearance; they are invariably Gram-negative and may possess peritrichously arranged flagella. All are obligate aerobes and grow best on nitrogen-deficient media, producing copious quantities of capsular slime if glucose is present as a carbon source. Three species are generally recognised, *A. chroococcum*, *A. vinelandii* and *A. agile*; a fourth species, *A. indicus*, has been transferred to the genus *Beijerinckia*[26].

These species occur in many different soils, though they are usually found in low numbers. In a recent survey of several agricultural soils, Brown, Burlingham and Jackson[54] found that, when present, their numbers varied from as few as twenty to as many as 8,000 per gram. Occasionally, large numbers were observed in rhizosphere soil but the increase was only about two- to threefold. In general, *Azotobacter* was found in soils with pH values from 5·9 to 8·4, although it was nearly always absent from the rhizosphere if the pH was below 6·5 and virtually never occurred naturally on root surfaces. These data substantiate results from many parts of the world[134, 420]. However, absence of *Azotobacter* from a soil is not only related to pH but also to a complex of interrelated factors such as inadequate phosphorus, calcium or trace elements, low humus content or lack of other energy-yielding materials, the presence of toxic substances produced in the soil and the degree of cultivation[420].

In soils where *Azotobacter* is not found, other free-living nitrogen-fixing bacteria may occur. *Beijerinckia* grows aerobically but only in acid, tropical soils, while *Clostridium* spp., which are intermediate between *Azotobacter* and *Beijerinckia* in their pH requirements, grow under anaerobic conditions in many soils. Certain *Pseudomonas* spp. may be responsible for nitrogen-fixation in acid forest soils.

Much of the interest taken in *Azotobacter* has arisen from the possibility of introducing it into soil in order to increase plant yields. Russian workers have reported increases in yield of cereals, tobacco, cotton and sugar beet of up to 20 per cent, following inoculation of soils with *Azotobacter* preparations. Preparations of *Azotobacter* for inoculation into soil may be made on a small scale, by harvesting two-day-old cultures of *Azotobacter* grown on Ashby's medium and mixing them with peat (about 40 mg bacteria per kg peat). After incubating this for several days, the peat is diluted until it contains about 10^8 cells per gram. Large-scale preparations are made by mixing $\frac{1}{2}$ ton of peat, 5 kg sugar or molasses, $\frac{1}{2}$-1 kg superphosphate, 10 kg chalk and *Azotobacter* cells from 25 Roux bottles. This mixture is incubated at 20-25°C in a layer 20-40 cm thick and then mixed with soil. Alternatively, aqueous suspensions or freeze-dried material may be applied to soil, the latter surviving for a longer period, although their main use is for soaking and spraying seeds prior to planting.

Many scientists are sceptical of the results obtained with these suspensions. They point out that because the organism is uncommon in soils (especially on root surfaces), and because there is insufficient energy-yielding material in the soil to support nitrogen-fixation on the scale required, introduction of *Azotobacter* could only be achieved by modification of the soil environment (which in itself would affect plant growth). Furthermore, the increased yields are not readily reproducible. Alexander and Wilson[12] showed that *Azotobacter* (the most efficient of the free-living nitrogen-fixing organisms) would have to oxidise 1 g of sugar in order to fix 5-20 mg of nitrogen. Translated to a field scale this means that 1,000 lb of organic matter would be required to fix 5-10 lb of nitrogen per acre per annum, or put another way, cell turnover rates of 50,000 per day would be needed to produce an annual increase of 2 lb of nitrogen per acre[6]. Such levels of activity are inconceivable.

Most of the inoculation experiments performed outside the Soviet Union have given negative results, although the scale on which these experiments has been carried out cannot compare with that of the Soviet work. A recent investigation of this problem has been made by British workers[55], and their results go some way towards substantiating certain of the Russian claims. They found that, although they inoculated *Azotobacter* on to seeds and roots and into soil, it only survived in the rhizosphere and never in root-free soil. The numbers present in the rhizosphere were governed by inoculum size and age. Thus a minimum inoculum of 10^4 bacteria per seed was needed for successful establishment, while encysted 14-day cultures survived better on seeds than two-day cultures if the seed was stored before sowing. They showed that the establishment of *Azotobacter* by seed treatment was due to its ability to migrate along the roots, keeping pace with root growth. This migration and possible multiplication was probably promoted by exudates from the young root, including various sugars. However, the cells did not seem to multiply on mature roots, although they were still carried passively in the moisture film around the roots. Migration seemed to be limited to the roots, for they were unable to detect much movement in soils without roots. Exudates from the roots were also responsible for induction of germination of *Azotobacter* cysts, suggesting that a phenomenon akin to fungistasis operated in soil (Chapter 9).

The effect of such inoculation was to increase crop yield, germination rates, root growth and speed of plant development, especially in tomato. These effects were usually observed during the early stages of growth when only small numbers of *Azotobacter* cells were present, suggesting that they were caused, not by fixation of large amounts of nitrogen, but by the production of small amounts of highly active growth-promoting substances. This possibility, together with suggestions that *Azotobacter* might inhibit plant pathogens and hence increase plant yields, is explored further later in this chapter.

These alternative explanations of the significance of *Azotobacter*, interesting as they may be, do not account for the accumulation of nitrogen in soils where other nitrogen-fixing organisms are absent. Increases in nitrogen content have been reported in soils planted with cereals, pasture grasses and trees where, in the absence of root-nodule bacteria and blue-green algae, nitrogen losses would be expected because of cropping, leaching and volatilisation. Possibly, *Azotobacter*, *Clostridium*, *Pseudomonas* and other less efficient nitrogen-fixing bacteria may be able to fix nitrogen in small quantities, contributing to a large cumulative effect, although it is still not possible to account for the source of energy for this process. Very little is known about rates of nitrogen-fixation in such conditions and until reliable information can be obtained by the use of radio-isotope techniques in the field, little progress will be made.

(*ii*) *Blue-green Algae* Unlike most other nitrogen-fixing organisms, blue-green algae flourish only in the very surface layers of the soil where light is available to them. Agriculturally, they are most important in rice paddies where De and Mandal[108] have calculated that they fix 13-70 lb of nitrogen per acre per year, or more if fertilisers or crop plants are present. Inoculation of paddies with blue-green algae has been attempted by Watanabe[548, 549], who reported nitrogen gains 20 per cent greater than those in uninoculated paddies, roughly equivalent to the addition of 26 lb per acre of ammonium sulphate fertiliser.

Stewart[483] has also investigated nitrogen-fixation by blue-green algae in natural environments. He found that the nitrogen fixed by these organisms in coastal soils could be transferred to plants growing in the same areas. Soils rich in *Nostoc* were exposed to an atmosphere containing ^{15}N in the laboratory for 21 days. Large quantities of

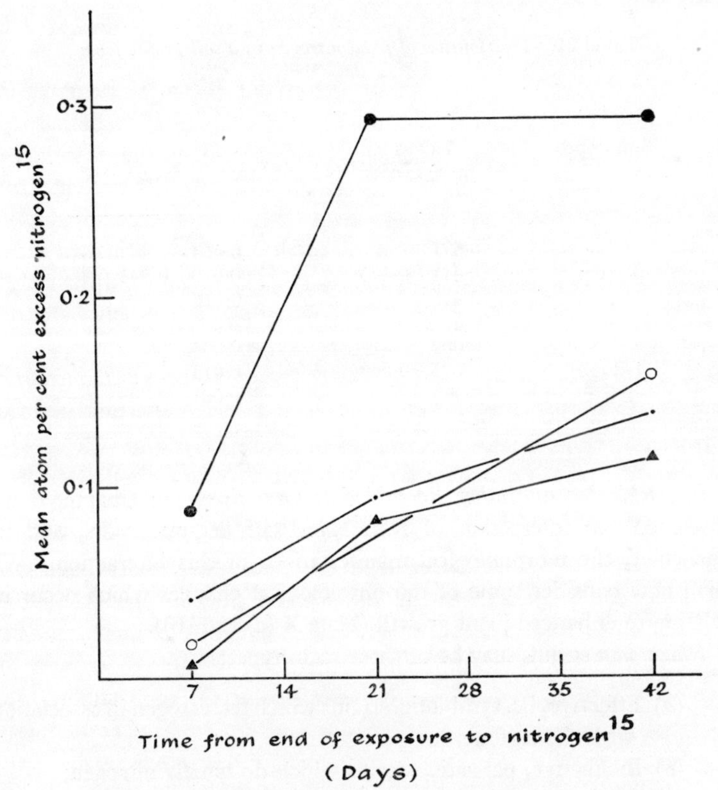

Fig. 23. *Uptake of previously fixed nitrogen by various plant species. At each harvest, all shoots were cropped* 0·5 *in. above ground-level. Experimental conditions as described in the text. Each point is the mean of triplicate determinations.* ● — ● *Bryum;* ○ — ○ *Glaux;* • — • *Suaeda;* ▲ — ▲ *Agrostis* (*Stewart*[483])
(*Reproduced by permission of W. D. P. Stewart and* Nature)

labelled nitrogen were taken up by *Nostoc* but labelled nitrogen was also found in higher plants that had grown in the soil during the same period. Figure 23 shows the rate of uptake of fixed nitrogen into various plants, after atmospheric nitrogen had been removed from the system. A species of the moss *Bryum* accumulated the nitrogen most rapidly.

Practically no fixed nitrogen was detected in the sub-surface soil, indicating that in the absence of leaching, the fixed nitrogen was not mobilised, or if it was, then it has been taken up rapidly by the roots of higher plants (Table 21).

Table 21. *Distribution of fixed nitrogen in a soil profile from a dune slack*[483]

Soil depth	7 days Atom % excess ^{15}N		21 days Atom % excess ^{15}N	
0-1 cm	0·571 0·411 0·436	0·473	0·493 0·670 0·616	0·593
1-2 cm	0·062 0·075 0·043	0·060	0·044 0·013 0·001	0·019

(*iii*) *Rhizobia and other Root Nodule Organisms* In Chapter 5 we discussed the interaction of *Rhizobium* with legume roots, and in particular, the morphological manifestations of this interaction. We shall now consider some of the physiological changes which occur in relation to enhanced plant growth (Plate X (a) and (b)).

Rhizobium strains may be of three main types:

(*a*) Effective, i.e. symbiotic, strains which fix nitrogen in association with the legume;
(*b*) Ineffective, parasitic, strains which do not fix nitrogen;
(*c*) Ineffective parasitic strains which do not fix nitrogen and which also prevent chlorophyll formation.

Plainly, the relative proportions of the different strains in the vicinity of plant roots could be important in determining the growth of the legume. In poor hill-pasture soils, the percentage of ineffective strains of *Rhizobium trifolii* may be high[214]; as a result, clover is difficult to establish in these soils. In certain instances, the proportions of the strains present in the soil can be altered by pre-inoculation of seeds with effective strains[391]. However, Vincent and Waters[530] showed that the proportions of the strains found in the nodules were not related to the proportions of the strains in the soil, and Means,

Johnson and Erdman[315] showed that a strain of *R. japonicum* comprising only 1·1 per cent of the soil inoculum was present in 85 per cent of the nodules formed under greenhouse conditions. Subsequently, Johnson, Means and Weber[239] have suggested that very heavy inoculation of highly competitive bacteria should be added to soil for successful introduction of a new strain. There are many examples of increased crop yield following inoculation. Norman[333] noted that inoculation increased seed yields of soyabeans by 31 per cent and Lynch and

Table 22. *Fixation of nitrogen by different legumes*[482]

Legume	Nitrogen fixed in lb per acre per year
Lucerne	158
White clover	133
Mixed legumes	112
Red clover	103
Sweet clover	93
Soyabean	84

Sears[283] found that in 89 out of 108 comparisons of inoculated and uninoculated soyabeans, increases in yield were obtained. Naturally, in soils where no rhizobia are present, pre-inoculation is essential if high yields are required during the first few years of cropping, but in other soils, where natural populations of effective rhizobia exist, pre-inoculation may be no more than a cheap insurance against crop failure (Ham, personal communication).

Not all nodulated legumes fix nitrogen to the same degree. Stewart[482] has summarised the data on the average amounts of nitrogen fixed by legumes and his results are reproduced in Table 22, from which it can be seen that white clover and lucerne are the most efficient.

It is not surprising that legumes are much richer in proteins than other plants. Whyte, Nilsson-Leissner and Trumble[558] found that the mean protein content of legumes was 13·8 per cent, compared with 5·3 per cent for grasses. Legume cultivation may be advantageous, therefore, since they may be ploughed back into the soil as a green manure to maintain and increase soil fertility; they may be grown in association with grasses to provide rich grazing pasture; and they may be used as a pure crop for hay or grain. Heard[199] investigated the effect of various legumes, grasses and grazing treatments on the growth of different crops planted in the same soil in the three subsequent years. He found that, compared with cocksfoot seed production leys, the residues of grazed leys containing lucerne gave rise to an extra 95 lb of nitrogen per acre taken up by test crops over three years (two crops of wheat and one of kale). White clover leys provided an extra 82 lb, grazed grass clover mixtures 45 lb and rye grass alone only 15 lb. The first wheat crop following grazed lucerne gave yields of 32 cwt grain per acre, compared with 23 cwt following rye grass. However, as Stewart[482] points out, introducing legumes into a crop rotation scheme may be difficult for an arable farmer since he cannot afford to let his land lie fallow for a whole year. For this reason, crops of legumes grown during the winter are becoming increasingly popular. In some instances, legume cultivation is highly profitable and soyabeans have now become the leading cash crop in the United States, exceeding traditional crops such as corn.

There is less information available on the amounts of nitrogen fixed by non-leguminous plants, the rate of addition of such fixed nitrogen to the soil or the transfer of this nitrogen to other plants. Amongst the information recently collected together by Becking[27] is the following. *Alnus* contributes 61·5-157 kg nitrogen per hectare per annum by leaf fall alone, while the afforestation of a tropical soil with *Casuarina* can result in an average net nitrogen fixation of 58·5 kg nitrogen per hectare per annum. Mature *Hippophae* trees may fix at least 69·5 kg nitrogen per hectare per annum in Britain. Growth of nodulated non-legumes may have beneficial effects on other plants. Thus *Ceanothus velutinus* improves the growth of *Pinus ponderosa* while *Ceanothus cordulatus* not only improves the growth of *Ribes roezli* but also increases the survival of *Ribes* seedlings following its growth. Similarly, *Ceanothus leucodermis* has been shown to improve the growth of succeeding crops of tomatoes.

The role of naturally occurring nodulated plants is not clear, but undoubtedly they are of importance in soils low in nitrogen. Legumes and plants like *Hippophae* are typically found as pioneer colonisers in dune and desert areas, while *Eleagnus* grows on eroded slopes and *Myrica* in acid, peaty soils. In the Arizona deserts, leguminous shrubs act as 'nurse' plants for other plants, including the saguaro cactus, *Carnegia gigantea*, during their early stages of growth by providing a supply of nitrogen and shade.

(b) Phosphate and Other Substances

It is clear that many mycorrhizal fungi affect plant growth by accumulating soluble nutrients from the soil and transferring them to the plant, and by synthesising growth substances of various sorts.

From the studies of Harley and his co-workers, we have good evidence that ectotrophic mycorrhizal fungi assist in the uptake of phosphate by plant roots. Phosphate from the soil is absorbed into the fungal sheath where much of it is accumulated. Only a small amount is passed to the root if the external supply is plentiful, but if the external supply is low, a good deal of accumulated phosphate is transferred. It is likely that the fungus traps and stores phosphate shortly after leaf fall when large amounts are present in the soil and provides the host with a steady supply during the rest of the year. There is some evidence that the uptake of other minerals is also enhanced and Harley and Wilson[188] showed that beech mycorrhiza vigorously absorb potassium ions. If nothing else, the infection leads to an increase in the absorptive surface area of the root presented to the soil.

Similar evidence exists for the participation of vesicular-arbuscular mycorrhiza in phosphate uptake for Gray and Gerdemann[153] found that infections by *Endogone fasciculata* enhanced uptake of ^{32}P by the host root from nutrient solutions and soil. Mosse[326] has also shown that mycorrhizal apple roots increased plant growth in sterile soil, possibly through the increased uptake of potassium, calcium, copper and iron and the decreased uptake of magnesium and manganese. Daft and Nicholson[101] found that the extent of growth stimulation in tomato, tobacco and maize plants depended on the level of inoculum of the fungus and the nutrient status of the environment. Marked stimulation of growth occurred in soils which were phosphate deficient, especially when the inoculum was high.

The role of the endotrophic mycorrhiza of orchids is somewhat different. If orchid seedlings are kept free of infection and grown in culture, they usually require a supply of sugars and vitamins. Infected plants do not need these supplements, suggesting that the fungus supplies the plant with these substances. Smith[456] showed that *Rhizoctonia solani* could compete successfully with saprophytes in soil for cellulose, which it broke down to sugars. Use of radioactive isotopes demonstrated that some of these were translocated by the fungus to orchid seedlings which it had infected. Fungal infection and the supply of carbohydrates and vitamins also promotes germination of orchid seeds.

Production of Growth-regulating Substances

Micro-organisms are prolific producers of growth substances and vitamins, i.e. substances which induce marked changes in plant growth when supplied in small quantities. The ability of micro-organisms to synthesise indole acetic acid was known long before its growth-regulating properties were suspected[216]. In a survey of 150 different fungi, bacteria and actinomycetes isolated from soil, Roberts and Roberts[397] found that 66 per cent would produce substances that provoked a growth response in *Avena* coleoptiles, and Brian[47] found that most of the 25 soil fungi he investigated could produce indole acetic acid. Unidentified growth factors have also been found in extracts from cells of the alga *Phormidium* and Shukla and Gupta[447] have shown that pre-treatment of rice seeds with such extracts can accelerate germination, promote shoot and root development and increase the protein content of the rice. Cook and Lochhead[92] found that almost 80 per cent of the bacteria they isolated from the root region of wheat and 37 per cent from root-free soil could synthesise one or more of nine vitamins (see Chapter 9). Several of these, e.g. thiamin and biotin, are known to increase plant growth.

Roots do not synthesise vitamins and under normal circumstances obtain adequate supplies from other parts of the plant[431]. This does not preclude the possibility of growth factors produced in the soil having additional effects on plant growth, for we know that roots can absorb growth regulators and hormones[211]. In general, it is not known whether soil micro-organisms can produce enough growth factors especially in the rhizosphere, where they may be degraded by other micro-organisms, although ectotrophic mycorrhizal fungi stimulate branching of short roots and prolong their life. This whole problem may be conveniently examined by further reference to *Azotobacter*.

We have already noted that increases in plant growth ascribed to the action of *Azotobacter* are capable of a variety of explanations. One of these explanations is that *Azotobacter* produces growth hormones in microbial composts which are then imbibed by the seed or in the rhizosphere after the seeds have germinated. Rasnizina[389] was the first to show that *Azotobacter* could form auxins and recently Brown, Jackson and Burlingham[55] detected 0·05–1·0 μg of indole acetic acid per ml of culture fluid. Vančura and Macura[522] found that while the effect of adding 1 ml of *Azotobacter* culture fluid to germinating seeds on agar was equivalent to the addition of 6 μg of indole acetic acid, such quantities were never found in the culture filtrates. Under natural soil conditions, it is possible that indole acetic acid is synthesised in greater quantities. Smalii and Bershova[455] suggested that auxin production by *Azotobacter* was greater in mixed culture and in pure cultures mixed with the metabolic products of *Bacillus mycoides*, *Agrobacterium radiobacter* and *Streptomyces coelicolor*, than in pure culture.

In the absence of conclusive proof of the involvement of indole acetic acid, the role of other growth regulators has been investigated. Vančura[521] found up to 20 μg/ml of gibberellic acid in 17-day-old cultures of *Azotobacter*. However, Brown *et al.*[55] were unable to obtain such high levels of production and concluded once more that the amounts produced were insufficient to account for the observed morphological effects. Five micrograms of gibberellic acid applied to seeds or roots were needed to produce the same effect as *Azotobacter* cultures. At least one other growth substance was detected in their experiments.

No information is available on the rate of growth substance formation in the rhizosphere so the role of these substances remains obscure. *Azotobacter* was selected for use as a plant inoculant because of its nitrogen fixing ability. If it turns out that its effect is due to growth hormone production, the possibility that other more efficient hormone producers might be used as inoculants will have to be investigated[10]. The ability to produce these substances is widespread and Katznelson and Cole[245] have reported the occurrence of gibberellin-like compounds in culture filtrates of bacteria and actinomycetes, including the common soil isolate *Arthrobacter globiformis*.

Growth regulators are also synthesised in large amounts by certain plant parasites and these may cause normal but exaggerated growth responses or loss of the typical organisation of the plant. *Fusarium moniliforme* (perfect stage *Gibberella fujikuroi*), which causes bakanae disease of rice, was the organism from which gibberellins were first isolated. Also, the action of *Agrobacterium tumefaciens*, the causal agent of crown gall disease, is thought to involve growth regulators, as yet uncharacterised but referred to as the tumour-inducing principle. *A. tumefaciens* can survive in soil for very long periods and some authors have suggested that it can live saprophytically. Survival of this parasite in soil is favoured by low temperatures, moist and alkaline conditions and fine soil texture[111].

The Solution of Plant Nutrients for Growth

It is well known that micro-organisms, in breaking down organic matter, release inorganic nutrients which may then be used for plant growth. Without the intervention of micro-organisms, organic matter would accumulate and nutrients would gradually become unavailable to plants. In areas where microbial activity is low, e.g. bog land, peat deposits may build up since the rate of synthesis of organic matter far exceeds its rate of breakdown. The whole of the saprophytic soil microflora participates in this process of nutrient release, perhaps the one undisputed example of a beneficial effect of saprophytes on plant growth. Also, as we saw earlier in this chapter, *Rhizoctonia*, the symbiont of orchid mycorrhiza, may break down carbohydrates and pass the resulting sugars into the host plant (p. 141).

The possibility that micro-organisms release from inorganic materials, nutrients which may benefit plant growth is rather more controversial, for nutrients that are released may be immobilised within the microbial cells. Different aspects of this problem may be illustrated by reference to the role of micro-organisms in attacking phosphates and silicates.

In young, raw soils and in the parent materials from which these soils are derived, phosphates and silicates are generally present in an insoluble form, often combined with iron and aluminium. It has been estimated that the average phosphate content of many soils is about 0·04 per cent. Kobus[254] has calculated that plants need approximately 25-100 kg of phosphorus (P_2O_5) per hectare per annum, most of which must be derived from the phosphate reserve and insoluble phosphate fertilisers. Silicates are also largely insoluble, but their breakdown by weathering or microbial action can lead to the liberation of potassium ions and other nutrients and to the production of clays[94]. It is not surprising, therefore, that some attention has been paid to the role of micro-organisms in dissolving these materials.

There is now clear evidence that micro-organisms produce organic acids, e.g. 2-keto gluconic acid (bacteria) and citric and oxalic acids (fungi), which dissolve these minerals at least in culture[555]. Phosphates resistant to these acids, e.g. iron and aluminium phosphates, may be dissolved by the action of microbially produced hydrogen sulphide[402, 465].

The greatest numbers of silicate and phosphate decomposers are found in soils with good supplies of organic matter where they comprise a substantial proportion of the microflora. Webley *et al.*[555] showed that many organisms from rocks and poor soils could attack various silicates (Table 23). Similarly, Kobus[254] estimated that 8-95 per cent

Table 23. *Attack of silicate minerals by soil micro-organisms*[555]

	Total no. of isolates	Percentage attacking minerals			
		Calcium silicate	Wollastonite	Magnesium silicate	Zinc silicate
Bacteria	265	83	57	65	not tested
Actinomycetes	39	87	38	46	not tested
Fungi	149	94	not tested	76	96

of the micro-organisms he studied in different soils would attack calcium diphosphate. Not surprisingly, increased numbers of phosphate dissolvers are found in the rhizosphere[147] and very large numbers on the root surface itself[281, 282]. However, no preferential stimulation of these organisms could be detected in these experiments, although preferential stimulation of organisms attacking organic phosphorus compounds has been noted in various grasses[159]. Of these compounds, phenolpthalein diphosphate, glycerophosphates and sodium phytate were attacked most frequently, lecithin and deoxyribonucleic acid least frequently. Despite the fact that micro-organisms can break down organic phosphates in culture, these compounds are often resistant to decomposition in the soil because they are adsorbed by clay minerals and sesquioxides. Indeed, substrates such as phytic acid may accumulate in chernozem soils which are particularly rich in organic phosphates. These organic phosphates may originate from micro-organisms for Duff, Webley and Scott[121] found that 17 per cent of the phosphate released from dicalcium phosphate was converted to organic phosphates by micro-organisms.

This immobilisation of phosphate by micro-organisms is well illustrated by some recent experiments with barley carried out by Barber and Loughman[19]. They showed that when the concentration of phosphate external to the root was low (below 0·5 p.p.m.), micro-organisms accumulated phosphate at the plants' expense. This was reflected in reduced transfer of phosphate to the shoots under non-sterile conditions and the increased amounts of phosphorus found in nucleic acids, phospholipids and phosphoproteins of the root surface micro-organisms. When these non-sterile root systems were transferred to phosphate-free solutions, more phosphate was lost than from corresponding sterile roots, due to its release from micro-organisms and their migration into the solution. Barber[18] consolidated this evidence in experiments showing the effects of pH on phosphate uptake by plants. He examined phosphate uptake at three different pH values, pH 4·0, 6·0 and 8·0, under sterile and non-sterile conditions. He found that in sterile conditions, although the amounts of phosphate transferred to the shoots varied, most being taken up at pH 6·0, the partition of phosphate into soluble, DNA, RNA and residual fractions was unaffected by changes in pH. In non-sterile conditions, the partition of the phosphate was affected, as with increasing pH, relatively less was found in the soluble fraction and more in the RNA fraction and less was transferred to the shoots. This change was almost certainly due to the intervention of micro-organisms which he showed were inhibited at low pH values. Not surprisingly, therefore, most phosphate was taken up by non-sterile root systems at pH 4·0. The interpretation of this data is complicated, since the ionic status of phosphate is pH dependent; at pH 4·0, 99 per cent of phosphate occurs as $H_2PO_4^-$ ions, while at pH 8·0, HPO_4^{--} ions predominate, only 13 per cent being present as $H_2PO_4^-$ ions.

Quite different results have been obtained by Rovira and Bowen[417], working with clover and tomato. They found that under non-sterile conditions, root growth was inhibited but more phosphate was taken up and transferred to the shoots. In tomato, 4·4 times as much ^{32}P was transferred to the shoots. Barber and Loughman[19] have suggested that differences in experimental procedure account for this disparity, but it could also be due to the use of different plants and to the different soils used to inoculate the roots. Plainly, we need to know a great deal more about a variety of plant microbe systems before we can make generalisations about the beneficial and harmful effects of the soil microflora in this process. All investigators are agreed that these results materially affect hypotheses put forward by physiologists to explain nutrient uptake by plants. In many instances, wholly inadequate precautions have been taken to exclude microbes from experimental systems.

The evidence presented here makes the value of applying bacterial fertilisers, such as phosphobacterin, to soil doubtful. Phosphobacterin consists of a compost containing *Bacillus megatherium phosphaticum* which mineralises organic phosphorus compounds. Its application to soil at a dosage of 5 kg per hectare is said to increase plant yields by about 10 per cent, but unfavourable soil conditions, e.g. desiccation, and prolonged laboratory culture of the bacterium may prevent these increases from being achieved. Cooper[94] reports that Russian workers have obtained strains, specially adapted to grow in the rhizosphere of wheat, maize, lucerne and clover, which release more phosphate than they accumulate. These strains also release ammonia from bound nitrogen so that their action on phosphates may not be the main reason for their beneficial action on plant growth.

Suppression of Plant Pathogens

Pathogenic micro-organisms are responsible for considerable losses of crop plants. Since many of the pathogens spend some period of their life in the soil, it has been suggested that the antagonistic properties of the soil saprophytes might be harnessed to suppress them and hence increase crop yields. The general phenomenon of antagonism between soil micro-organisms will be discussed in Chapter 9; here we shall deal only with some of the problems associated with the biological control of plant pathogens.

In common with many of the phenomena we have discussed in this chapter, inhibition of pathogenic organisms is easily demonstrated under controlled laboratory conditions but is difficult to prove in the field. However, there is much circumstantial evidence indicating the importance of such interactions. Typical of this evidence is that recently produced by Huber and Anderson[219] who showed that when *Xanthomonas* cells and *Fusarium solani*, the causal organism of bean root rot, were grown in mixed culture, it resulted in massive hyphal colonisation by the bacteria, agglutination of the hyphal contents, production of a reddish purple pigment in the hyphae, and death. A series of isolation experiments showed that growth of *Xanthomonas* in the soil was favoured by corn but not by barley. Also the severity of bean root rot was less in soil previously planted with corn. They suggested that the association of these results pointed to the controlling action of the bacterium, but increases in numbers of *Xanthomonas* cells and control of root rot may both result from some other change in the soil environment induced by the growth of corn. The evidence for microbial suppression of the pathogen is strong in this instance but not conclusive. These experiments underline the necessity for changing the soil environment to favour the growth of the antagonist, for it is unrealistic to expect an organism, natural or introduced, to multiply and control disease without such alterations. Changes in the environment can be brought about by green manuring, addition of chitin or laminarin, or by addition of various minerals. An early example of disease control brought about in this way was the use of sulphur to control potato scab[307]. The production of acid from the sulphur by sulphur-oxidising organisms suppressed the growth of *Streptomyces scabies*.

Bacteria are not the only micro-organisms credited with an ability to suppress pathogens. Stevenson[478] has shown that antibiotic-producing actinomycetes decreased root damage to wheat seedlings by *Helminthosporium sativum* when soil had been inoculated with these organisms. Timonin[510] also pointed out that the fungus *Trichoderma viride* could parasitise the hyphae of this and other pathogens, and Simmonds[448] found that some or all of the seed coat microflora of wheat gave protection against *Helminthosporium* infection.

Mechanisms whereby resistant plants might stimulate the growth of antagonists have been suggested. Timonin[510] associated the development of disease resistance of flax with the occurrence of linamarin which, when hydrolysed, yields hydrogen cyanide. Pathogens such as *Fusarium* and *Helminthosporium* are suppressed by hydrogen cyanide while antagonists like *Trichoderma viride* are immune. Clearly the release of hydrogen cyanide must be carefully controlled for excess production might inhibit plant growth. As we shall see later, Patrick[356] has invoked the release of hydrogen cyanide from 'dead' roots of peach trees as an explanation of peach tree replant problems.

Production of Phytotoxic Substances

It has been suggested that many soil-borne plant parasites and free-living micro-organisms may produce toxins or release toxins from plant residues in the soil which could inhibit growth.

The list of potentially phytotoxic chemicals that can be produced by free-living micro-organisms is impressive and includes organic and amino acids, antibiotics, gibberellins, nitrite, carbon dioxide and hydrogen sulphide. In addition, amino acids, alkaloids, oils, coumarin, lactones, phenols and aldehydes, organic acids, ammonia, hydrogen cyanide, mustard oils, scolopetin and other substances can all be released from soil organic matter by micro-organisms[294]. Before a reaction can be accredited to one of these substances under natural conditions, several criteria must be fulfilled.

(*i*) A potentially phytotoxic substance must be present in an effective concentration in the soil. Often only small quantities are required to inhibit growth, making them so difficult to detect that no one has yet succeeded in extracting a chemically defined substance of proven biotic origin in effective quantities from soil[357]. It is, of course, possible that substances present in the soil in ineffective concentrations become concentrated after absorption by plants. A further complication is that substances which are toxic at one concentration become stimulatory at another, e.g. gibberellins, ammonia, etc.

(*ii*) The substance must be taken up by the plant. Roots are quite capable of absorbing large molecules either by diffusion or active uptake and Winter and Willeke[569] have shown that antibiotics can be taken up and translocated to various parts of a plant. Larger molecules, e.g. enzymes and viruses, may also enter plant roots, possibly by ingestion through pinocytosis[299, 328].

(*iii*) The phytotoxin must persist within the plant or, if it is degraded, give rise to further toxic material which can produce the original symptoms.

Diagnosis of phytotoxicity may be complicated because of difficulties in distinguishing the symptoms from those of mineral deficiencies or onset of unfavourable environmental conditions. These similarities may be further complicated since inorganic materials can enhance or repress the action of a toxin. Thus iron has been shown to increase the potency of toxins produced by pathogenic fusaria[428].

Phytotoxins produced by free-living organisms are mainly associated with heavy, highly organic, poorly aerated soils with a high water content found in cool, temperate regions[292], conditions which themselves are inimical to growth. Under anaerobic conditions, intermediate breakdown products of cellulose and lignin, e.g. organic acids, will form. Toxin production seems to be associated with the early stages of decomposition. Harris[189] studied the production of phytotoxins during decomposition of wheat straw and showed that the decomposition of recently incorporated stubble had deleterious effects on subsequent cereal growth, especially during the first two to three weeks.

The symptoms induced in plants by phytotoxins are diverse and include inhibition of germination, reduction of root and shoot growth, wilting and increased susceptibility to diseases. However, there are very few examples of these effects being positively linked with phytotoxicity in the field. A probable case of phytotoxicity is the disease known as frenching of tobacco plants. The plants develop large numbers of axillary shoots and many abnormal leaves. Steinberg[476] has suggested that this is caused by a toxin produced by *Bacillus cereus*, which causes excessive accumulation of isoleucine and other free amino acids in the plant, which are also known to induce frenching. McCalla and Haskins[294] have reviewed some other likely examples, including peach tree replant and citrus replant problems and stunting of corn following stubble mulching. Patrick[356] showed that peach roots produce the non-toxic glucoside amygdalin; microbial action converts this to glucose, benzaldehyde and hydrogen cyanide which may inhibit root growth of new trees. Similar problems, encountered when *Citrus* trees are replanted on old *Citrus* land, are probably due to the presence of root pathogens and toxins produced in other ways, but the presence of microbially produced phytotoxins cannot be ruled out[306].

Phytotoxins are also thought to be produced by a number of plant parasites which exist in the soil for some time, although there are relatively few examples of such substances which have been isolated in chemically pure form. Amongst the best known ones are lycomarasmin and fusaric acid which are produced by *Fusarium oxysporum* f. *lycopersici* and which respectively cause necrosis of the leaf and browning of the vascular tissue in tomatoes. Lycomarasmin is a peptide, similar in structure to the growth factor strepogenin, and probably acts as an inhibitory structural analogue. Fusaric acid (5-butylpicolinic acid) and lycomarasmin are also able to chelate heavy metals, and this may be important since it has been observed that EDTA is able to induce similar symptoms. These toxins are relatively unspecific, but there are some examples of highly specific toxins produced by soil-borne parasites, e.g. the tricyclic secondary amine victoxinine produced by *Helminthosporium victoriae* which attacks the basal region of oat plants. It affects only those varieties which bear the Victoria gene, and susceptibility is inherited in a simple 3:1 Mendelian ratio. A review of these and other toxins is given by Braun and Pringle[43].

Production of Enzymes

There is little evidence that enzymes produced in the soil outside the host plant can attack plant tissues, although we know that enzymes formed by pathogens inside the host may affect plant growth considerably[133]. It is known that enzymes can be taken up by roots so that if they were produced in large quantities by rhizosphere or root surface organisms, some effect might be expected. Fåhraeus and Ljunggren[130] have suggested that polygalacturonase formation by the root might be stimulated by the presence of rhizobia near root hairs which could modify root-hair structure and allow infection (see also p. 122). Many different enzymes have been reported from so-called root exudates but it is not clear whether they are of plant or microbial origin. They include invertase, amylase, protease, phosphatase, nuclease and cellulase[419].

Once inside a host plant, the production of enzymes by pathogenic fungi and bacteria is well known. Pectic enzymes cause the maceration and killing of parenchymatous cells in the soft-rot diseases and they have been implicated in some of the vascular wilts. Soil organisms which can cause soft-rots include the bacterium *Erwinia* and fungi such as *Pythium debaryanum* and *Sclerotinia sclerotiorum*. Vascular wilts may be caused by species of *Fusarium* and *Verticillium*.

The effects produced by these pathogens cannot always be explained solely by the production of pectinase and it is likely that many enzymes including proteinases and cellulase are also involved. The evidence for this is fragmentary and more critical work will be needed to establish their precise role[574].

Other Competitive Interactions between Plants and Micro-organisms

Microbes compete efficiently with plant roots for mineral nutrients since their metabolism is more intense. More ATP is generated per unit weight of tissue in a given time, providing a supply of energy for ion uptake[329]. It has been suggested that competition will be even more marked in anaerobic environments, because many micro-organisms are still able to generate energy whilst roots are inhibited. Thus denitrifying bacteria can utilise nitrate as an electron acceptor instead of oxygen during respiration. In this way, they not only remove nitrate from the root environment but probably compete for other nutrients as well.

Denitrification may be defined as a process whereby nitrate, through the agency of nitrate reductase and cytochromes, is reduced to nitrite, nitrogen or nitrous oxide during respiration. It can be carried out by several species of *Pseudomonas*, *Micrococcus* and *Bacillus*[450]. *Thiobacillus denitrificans* utilises nitrate as a hydrogen acceptor for the oxidation of thiosulphate, bringing about the same results. Denitrification is inhibited by oxygen, probably because oxygen competes with nitrate for the role of electron acceptor. Consequently, highest denitrification rates are found in waterlogged soils, especially those supplied with nitrate fertilisers or in which nitrification has recently taken place, whilst the lowest rates occur in well aerated soils[45]. Estimates of the rate of loss of nitrates from soil due to denitrifying organisms are subject to error and although figures of 20 lb nitrogen per acre per year are thought to be reasonable, Russell[425] has pointed out that it is difficult to prove whether these losses are due to microbes or plants (or presumably to leaching).

Not only is nitrate removed from the soil during denitrification but toxic nitrite is formed. It has been shown that nitrite toxicity can be an important factor in rice cultivation if nitrate fertilisers are used, but

Table 24. *Numbers of organisms* ($\times 10^6$) *per g soil under permanent grassland*[571]

Micro-organisms	Conditions of incubation: Anaerobic+ nitrate	Anaerobic. No nitrate	Aerobic. No nitrate
Total number	85	8	32
Bacteria	85	8	24
Total denitrifiers	73	6	4
Pseudomonas strain 241	39	0	0
Achromobacter strain 181	12	0	0
Bacillus macerans	12	0	0
Bacillus cereus	2	5	3
Other denitrifiers	8	1	1
Other bacteria	12	2	20

the adsorption of nitrite on clay minerals and its reaction with ammonia at acid pH values may partly offset this toxicity in some soils.

Woldendorp[571] has shown that denitrification may be particularly active around living roots. He compared the effect of both living and dead roots on denitrification and found that whilst only 9 per cent of added nitrate nitrogen was lost from soil containing dead roots, between 15 and 37 per cent disappeared from soil containing living roots. He postulated that this was due to low oxygen tensions in the rhizosphere, initiated through intense microbial respiration, the presence of suitable hydrogen donors and the large numbers of denitrifying bacteria present.

In the soils that Woldendorp[571] studied, the most prominent denitrifying bacteria were *Bacillus* spp., especially *B. cereus*. When nitrate fertilisers were added, these forms were replaced by *Pseudomonas*, *Achromobacter* and *Bacillus macerans* (Table 24). He concluded that the indigenous denitrifiers were present, not as a result of their denitrifying activities, but because of some other property they possessed. Consequently, no relationship between the original number of denitrifying bacteria in the soil and the denitrifying capacity of that soil could be established.

Denitrification undoubtedly contributes to the recovery of less than 60 per cent of added fertiliser nitrogen from grass tops[571]. The rest is either immobilised in the roots, root-surface micro-organisms and soil organic matter, or is leached out or volatilised. Parr[354] in a review considering methods for increasing the efficiency of nitrogen fertilisers, has stressed the need for controlling the release of nitrogen into the soil so that an excess, which would promote competitive reactions, does not build up. This might be achieved by using fertilisers which are recalcitrant and attacked only by very specific organisms that might be controlled. An effective alternative would be to coat fertilisers with different amounts of recalcitrant substances which would result in slow release of nitrogen over a long time period (see also p. 160).

Effect of Microbes on Soil Structure

A good soil structure is essential to plant growth, as it promotes gas exchange, penetration of water and roots and resistance to erosion. Evidently the best structure is achieved by the aggregation of soil particles into crumbs. These aggregates have been defined by Griffiths[174] as 'naturally occurring clusters or groups of soil particles in which the forces binding the particles together are much stronger than the forces between adjacent aggregates'. The optimum size for these aggregates is between 1·0 and 5·0 mm in diameter[424]. Greenwood[162] showed that 30 per cent of the volume of 7·0 mm diameter crumbs was anaerobic, whilst 35·0 mm diameter crumbs were 80 per cent anaerobic. Large crumbs also result in large pore spaces in which roots become desiccated. Small crumbs are accompanied by small pores in which gas exchange is prevented, again resulting in the development of anaerobic conditions. Currie[100] has pointed out that anaerobic conditions may lead to an upset balance in nutrient uptake by plants, a decrease in permeability of roots to water and an inhibition of root elongation. Also, ferrous and manganous ions increase in concentration in the soil while nitrates and sulphates are reduced to useless and toxic forms. Increases in carbon dioxide concentration may cause the pH to drop in poorly buffered soils, and products of anaerobic respiration may be phytotoxic.

Many factors are involved in the formation and stabilisation of soil aggregates, quite apart from micro-organisms. Formation may be brought about by flocculation and sedimentation of colloids, drying of a wet soil, tillage, root pressure, earthworms and moles. Stabilisation of these aggregates may be caused by inorganic cementation with iron and aluminium, clay, organic matter and earthworm slime[293]. In normal, arable loams, clays are the principal aggregating agent unless more than 2 per cent organic matter is present.

Two mechanisms have been suggested to explain the involvement of micro-organisms in aggregate formation. Firstly, filaments may bind material together, and secondly, polysaccharides and other adhesive agents produced by micro-organisms may glue the particles together. These two mechanisms may not be clearly separable for decomposition of fungal hyphae may lead to gum formation. The best examples of filaments binding particles together occur in sandy soils. Bond[37] observed this in Australian soils and we have seen similar aggregates bound by basidiomycete mycelium in British dune soils (Plate X (c)). Bond suggested that mycelium involved in binding might have a sticky surface.

The role of microbial gums in aggregate formation has been investigated by Mehta, *et al.*[317]. They made artificial aggregates with microbial gums and showed that these aggregates could be disintegrated by treatment with periodic acid. On the other hand, naturally occurring aggregates were not affected by periodic acid, suggesting that the binding was due to other forces. Greenland, Lindstrom and Quirk[160] re-investigated this problem with a wide range of Australian soils and showed that whilst they obtained similar results with old pasture soils and those high in organic matter, crumbs in young pastures and cultivated soils could be disintegrated with periodic acid. This suggests that different mechanisms may operate in different soils. Harris *et al.*[190] have criticised this interpretation on the grounds that materials other than polysaccharides might be hydrolysed by periodic acid in an environment as complicated as the soil.

Under experimental conditions, aggregation of particles by micro-organisms can be achieved only after the addition of an energy source such as plant residues or sucrose. Waksman and Martin[537] showed that such aggregation was not due to the organic matter itself, for it had no effect when added to sterile soil. Aggregates are formed following amendment with sucrose under both aerobic and anaerobic conditions, but Harris *et al.*[190] found that these aggregates could only be maintained under anaerobic conditions. They postulated that anaerobes were unable to break down the aggregating agents. In sterile soil, McCalla and Haskins[293] showed, by inoculation experiments, that organisms varied in their ability to produce gums. *Stachybotrys atra* was one of the most efficient, but attempts to introduce this fungus into non-sterile soil failed and although inoculation after fumigation of the soil was possible, the degree of sterility required to allow establishment could not be achieved without undesirable side-effects.

In some circumstances, production of microbial gums might have deleterious effects on soil. Pores can be blocked up, slowing down percolation of water, and coating of particles with non-wettable gums leads to increased water run-off[293]. Griffiths[174] has suggested that non-wettable substances like lipids may be responsible for aggregate stability as estimated by normal wet-sieving techniques. Only small amounts of lipid, coating the soil pores, would be needed to waterproof the aggregates, which might enable other substances, such as polysaccharides present in small amounts, to cement the soil particles together.

8

Autotrophic Micro-organisms in Soil

Not all micro-organisms use organic matter as a major source of energy and nutrients for growth. Many are able to use carbon dioxide as a sole source of carbon and incorporate it into their cells with energy obtained either from the oxidation of inorganic substances or photochemical reactions. Organisms which carry out such reactions are said to live autotrophically. The process of carbon dioxide fixation is expressed in the following equation:

$$2CO_2 + \underset{\text{reducing power}}{4(H)} + \text{energy} \longrightarrow \underset{\text{carbohydrate}}{2(CH_2O)} + 2(O)$$

It follows from this equation that autotrophs require not only a source of energy for growth but also a source of reducing power to convert carbon dioxide to carbohydrates. It is the reactions providing this reducing power that account for the importance of autotrophs in the soil, for the carbon dioxide that they fix is negligible when compared with that incorporated by higher plants. Photo-autotrophs obtain their reducing power from the oxidation of inorganic substances and their energy from light: they include the algae and the photosynthetic bacteria. Chemo-autotrophs obtain both reducing power and energy from the oxidation of inorganic substances: they include some bacteria and a few actinomycetes[244]. It has generally been assumed that the autotrophs are the chief agents of the oxidation processes we shall describe, but this may not be so, for some heterotrophic bacteria, actinomycetes and fungi which are more numerous in soil can carry out the same reactions.

Little is known about the ecology of autotrophic organisms in soil because we lack reliable information on their physiological and biochemical versatility[430]. Only recently has it become evident that many autotrophs, even those previously regarded as strict autotrophs, can use some forms of organic matter, either as an energy source or as a carbon source and sometimes as both. In addition, it is known that many organisms can live both autotrophically and heterotrophically and this has considerable ecological implications since it extends the potential range of environments in which these organisms can grow.

The amount and type of information that we have on soil autotrophs is related to the degree of their supposed significance in soil. It is convenient, therefore, to treat separately each process in which they participate.

Nitrification

(*a*) *Types of Nitrifying Organism*

Nitrification is a process whereby ammonium nitrogen is converted to nitrite and nitrate nitrogen. This conversion can be brought about directly by light or during the metabolism of heterotrophs, e.g. *Aspergillus flavus*[432], but the activity of two autotrophic bacteria, *Nitrosomonas* and *Nitrobacter*, is probably the most important. Both these bacteria are small, Gram-negative, polar flagellate rods; *Nitrosomonas* oxidises ammonia to nitrite and *Nitrobacter* converts nitrite to nitrate. These reactions yield energy and a source of reducing power for the incorporation of carbon dioxide into the cell.

$$NH_4^+ + 1\tfrac{1}{2}\ O_2 \longrightarrow NO_2 + 2H^+ + H_2O + 66\ \text{kcal}$$
$$NO_2^- + 1\tfrac{1}{2}\ O_2 \longrightarrow NO_3 + 17{\cdot}5\ \text{kcal}$$

Several similar nitrifying organisms, e.g. *Nitrosococcus*, *Nitrosocystis*, and *Nitrospira*, have been described but very little is known about them. Some may be cultural variants of the more familiar forms or cultures contaminated with heterotrophic bacteria. Contamination of cultures with heterotrophs is one of the main difficulties encountered in isolating and working with nitrifying bacteria. If small traces of organic matter are produced by them, heterotrophs develop which, in turn, may stimulate the growth of the nitrifiers through the production of pyruvate[85]. The best isolation media are liquid[11] or ones solidified with silica gel or purified agar[461].

(*b*) *Rates of Growth of Nitrifying Bacteria*

Although nitrifying bacteria grow rather slowly, large quantities of ammonium nitrogen can be converted to nitrate in soil. Broadbent, Tyler and Hill[49] showed that between 7 and 88 lb of nitrate nitrogen per acre per day could be formed in loam and clay soils treated with various ammonium fertilisers (100-1,600 lb N/acre). Morrill and

Dawson[324] found that the generation time of *Nitrosomonas* was half that of *Nitrobacter* at pH values below 7·2 in soils perfused with ammonium and nitrite nitrogen respectively. This means that nitrite was removed as soon as it was formed, a fact of some importance since nitrite is phytotoxic. Generation times were pH dependent and varied from 100 hr at pH 6·2 to 38 hr at pH 7·6 for *Nitrosomonas*, and from 58 hr at pH 6·2 to 21 hr at pH 6·6 and above for *Nitrobacter*. The generation times were calculated from the rates of conversion of ammonium and nitrite nitrogen and assume that the relation between generation time and disappearance of substrate is the same in soil as it is in culture. It is known that laboratory cultures grow more quickly, as Skinner and Walker[452] found generation times of 8–16 hr for *Nitrosomonas* while Tobback and Laudelout[512] showed that of *Nitrobacter* to be 19 hr. The value for *Nitrosococcus* is 12 hr[460]. These figures are not truly comparable, however, because different cultural methods were used in each case.

(c) The Effect of Soil Reaction

Since pH affects the generation time of nitrifying bacteria, it is not surprising that it can influence both their occurrence and activity in soil. Indeed, to encourage nitrification it is often necessary to add lime to soil to neutralise acids produced during the process. Chase, Corke and Robinson[78] showed that liming poorly buffered soils increased the pH from 5·7 to 6·3 and nitrification by 10 per cent; further study showed that liming soils in the pH range 4·2 to 6·2 always increased nitrification. Numbers of *Nitrosomonas* cells in acid forest soils under *Acer* could be increased from 260 to 11,600 per gram and to 32,900 if phosphate was also added. Similar increases were recorded for *Nitrobacter*. It is not surprising, therefore, that the naturally occurring soils with the largest populations of nitrifying bacteria are alkaline[567].

The relationship of soil reaction to nitrification is more complicated than the above results suggest. Nitrification in culture is minimal below pH 5·0 and yet nitrifiers can be isolated from, and nitrates found in, acid soils with lower pH values. Presumably, strains with different pH requirements exist but these have never been isolated and the possibility that more alkaline niches occur in acid soils should not be ignored (see p. 44).

(d) The Effect of Soil Moisture and Aeration

Dommergues[117] showed that nitrification would not occur in soils with pF values greater than 4·3-4·7. Later, Miller and Johnson[323] obtained much the same results and also showed that maximum rates occurred at pF values of 2·2-2·7.

Water content affects oxygen concentration in a soil and more especially the rates of diffusion of oxygen (Chapter 2). Nitrifying bacteria are obligate aerobes and in culture oxygen deficiency soon limits growth[452]. Recent evidence suggests that they can still grow in low concentrations of oxygen. Greenwood[162] showed that nitrification was inhibited only when the oxygen concentration fell below 5×10^{-6} M and concluded that this was because nitrifying bacteria possess cytochromes with a very high affinity for oxygen. When the oxygen concentration fell below the critical level, nitrification stopped and was replaced by denitrification brought about by anaerobic bacteria (Chapter 7). Naturally, denitrification can only take place if the nitrates diffuse from the aerobic into the anaerobic zones of the soil and this is favoured if the interface between the two zones is extensive. This occurs in well aggregated soils where the centre of the aggregates are anaerobic and the outsides aerobic (p. 153). If only small aggregates are present, then no anaerobic zones will occur and nitrification will proceed rapidly[439].

(*e*) *The Effect of Organic Matter and Ammonia*

The results of studies on the effects of organic matter on nitrification have been contradictory. Early workers suggested that organic matter inhibited nitrification, but the rapidity of the process in manure heaps indicated that this was improbable and, furthermore, recent biochemical studies have suggested that nitrifying organisms can actually metabolise some organic substances. Van Gool and Laudelout[523] found that formate could be metabolised but that this had very little effect on the fixation of carbon dioxide. It seems likely that energy released from its oxidation is stored in the cell and serves as an energy reserve for maintenance of the cells under unfavourable conditions.

Nitrification, by definition, depends upon an adequate supply of ammonium and nitrite ions and many of the factors which favour it also stimulate ammonia production from organic matter. Consequently, nitrification may proceed until all the readily available ammonia yielding material has been exhausted. Seifert[438] found that nitrification rates in soils incubated at 20°C fell rapidly, whilst in those incubated at 2°C, slower nitrification occurred which was maintained for a much longer period. He suggested that ammonia had been exhausted at the higher temperature. Lack of ammonium nitrogen has also been suggested, as the cause of low rates of nitrification in grassland soils, by Robinson[398] who showed that some grasses can remove ammonium ions from soil very rapidly and therefore compete directly with nitrifying bacteria.

(f) *Significance of Nitrification in Soil*

Most crop plants use nitrate in preference to ammonium ions as a source of nitrogen. Consequently, nitrification has been regarded as a process to be encouraged in fertile soils. Unfortunately, nitrate ions are not readily adsorbed by soil particles and may be rapidly leached

Table 25. *Effect of a field application of N-serve on winter loss of nitrogen from a Canadian soil* [78]

Soil treatment Nov. 1964 (per acre)	Mineral nitrogen, April 1965 (lb/acre)		
	Ammonium	Nitrate	Total
Control	7	8	15
N-serve (10 lb)	17	11	28
Urea (60 lb N)	19	14	33
Urea + N-serve	50	12	62

away in the soil drainage water (see p. 152) and, as we have seen, nitrification may be followed by denitrification which leads to further nitrogen loss. Therefore, in order to conserve soil nitrogen it may be necessary to control or even prevent nitrification in certain circumstances. This might be achieved by the use of fertilisers which release ammonium nitrogen slowly, e.g. urea formaldehyde compounds and crotonaldehydes. Nitrate would also be formed slowly, but used rapidly by the plants, thus reducing losses of excessive amounts of nitrogen. Chase, Corke and Robinson[78] have also suggested that plant breeders might produce crops which could tolerate and utilise ammonium nitrogen, so that if a selective inhibitor of nitrification could be developed, significant quantities of soil nitrogen would be conserved. It is known that several plants, including grasses, spurges and sunflowers, produce antimicrobial substances such as chlorogenic acid and gallotannins, which suppress nitrification[40, 395, 484]. Hesse[206] also found that the amino acid methionine, present in some tree leaves, inhibited nitrification. The action of all these substances is unspecific and they may affect other soil micro-organisms, but Goring[151] found that 2-chloro-6 (trichloro-methyl) pyridine, now known as N-serve, does inhibit nitrification specifically. If it is applied at the rate of 0·4 lb per acre, nitrification is inhibited for a few weeks[78]. Applications of 10-20 lb per acre are effective for 4 months and Table 25 shows the dramatic effect on the nitrogen level of the soil in the spring when this compound was applied during the preceding autumn. Unfortunately, such treatments are uneconomic.

Sulphur Oxidation

(*a*) *Sulphur in Soil*

Sulphur occurs in many compounds which together constitute 0·01-0·05 per cent of the soil mass[473]. It has been suggested that up to 25 per cent of this sulphur is in the form of sulphate, although in soils of humid areas much of the sulphur is incorporated in organic matter[481]. Unbound sulphates are readily leached from the soil but under dry conditions sulphate may accumulate in the form of gypsum in a definite layer, some distance from the soil surface (p. 49). Recently, Bremner (cited by Burns[65]) has suggested that more sulphate may be present than suspected, bound to organic matter and therefore not detectable by the usual analytical procedures.

Sulphur deficiencies are now much more common in soil than in former years, due to the increasing use of sulphur-free fertilisers, the reduction in atmospheric sulphur pollution from coal and oil and the increasing crop yields that are being realised[65]. Stotzky and Norman[491] showed that micro-organisms are affected by these deficiencies: apart from nitrogen and phosphorus, lack of sulphur was the only mineral deficiency which could limit microbial growth and the effects were usually evident before crop plants were affected. Added amounts of sulphur, as small as 0·32-1·6 mg per 100 grams of soil, increased markedly the rate of glucose oxidation by micro-organisms.

The most readily available form of sulphur for plants and heterotrophic micro-organisms is sulphate. However, sulphur in plant tissue and organic matter is in a reduced form which is released very slowly during organic matter decomposition. A variety of reduced, inorganic sulphur compounds are also found in soil, e.g. sulphides, polysulphides, etc., especially under waterlogged conditions. The organisms that can oxidise reduced sulphur are therefore very important for the maintenance of soil fertility.

(*b*) *Types of Sulphur-oxidising Organisms*

Sulphur oxidation is brought about by heterotrophic bacteria, actinomycetes and fungi[177] and by a variety of autotrophic bacteria. The latter are of three main types:

(*i*) Photosynthetic bacteria, e.g. *Chromatium* and *Chlorobium*;
(*ii*) Filamentous forms of uncertain physiology, e.g. *Beggiatoa* and *Thiothrix*;
(*iii*) Chemosynthetic, polar flagellate rods belonging to the genus *Thiobacillus*.

Photosynthetic bacteria are aquatic and are found most frequently growing anaerobically at mud/water interfaces where they obtain light. Often they grow under mats of algae which produce reducing substances and thus provide an anaerobic environment. *Beggiatoa* and *Thiothrix*, which are also aquatic forms, are almost certainly heterotrophs, although recent work suggests that they do obtain some energy from the oxidation of hydrogen sulphide in addition to that obtained from organic substances[437].

The only group of autotrophic sulphur-oxidising bacteria that are typical of soil are the thiobacilli. It was once thought that they were strict autotrophs but it is now known that some of them take up organic compounds and metabolise them[238]. They attack sulphur itself and a variety of sulphur compounds including sulphides, thiosulphates and tetrathionates, releasing energy and forming highly acidic substances. Two of the conversions brought about by these bacteria are given below:

$$2S + 3O_2 + 2H_2O \longrightarrow 2H_2SO_4 \text{ (\textit{T. thiooxidans})}$$
$$5S + 6KNO_3 + 2H_2O \longrightarrow K_2SO_4 + 4KHSO_4 + 3N_2 \text{ (\textit{T. denitrificans})}.$$

During the oxidation processes, energy is liberated as follows:

$$SH^- \underset{\substack{40\\ \text{kcal}}}{\longrightarrow} S \underset{\substack{15\\ \text{kcal}}}{\longrightarrow} S_2O_3^{--} \underset{\substack{5\\ \text{kcal}}}{\longrightarrow} S_4O_6^{--} \underset{\substack{100\\ \text{kcal}}}{\longrightarrow} SO_4^{--}$$

Some thiobacilli also obtain energy from the oxidation of ferrous iron to ferric iron (11·3 kcal), e.g. *T. ferrooxidans*, while *T. denitrificans* is unique since it uses nitrate rather than oxygen as a terminal electron acceptor during the oxidation of sulphur; consequently it is anaerobic and brings about denitrification.

(c) *Activities of Sulphur-oxidising Bacteria in Soil*

In most soils there are not many *Thiobacillus* cells, but these increase rapidly in number when sulphur is added[325]; increases from 10 to 200,000 cells per gram have been recorded. Sulphur is added to soil for a variety of reasons. For example, Lipman and MacLean[273] found that the acid produced from sulphur oxidation caused the solution of phosphate fertilisers, making them more available for plant growth, while Kittams and Attoe[252] have shown that the simultaneous addition of sulphur and rock phosphate to soil is equivalent to adding superphosphate. Furthermore, increasing the acidity by adding sulphur improves the structure of black alkaline soils by flocculating the colloidal materials. It can also suppress the growth of plant pathogens, e.g. *Streptomyces scabies*[307].

If excess sulphur is present in soil, sulphur-oxidising bacteria may have deleterious effects. Normand and Colmer[334] found that *T. ferrooxidans* increased the acidity of sulphur-rich, roadside soils, preventing the establishment of grass. Sulphur-oxidising bacteria have also been shown to produce acid from the copper sulphate/sulphur sprays used in vineyards; as the soils become acid, the copper becomes toxic to the vines[369].

With increasing demand for agricultural land, extensive areas of estuarine mud and mangrove swamp are being reclaimed. These muds are anaerobic and contain large amounts of sulphur, often in the form of polysulphides. Hart[192] showed that during reclamation the water content of the mud drops and at the 50 per cent level, the pH also starts to fall due to the production of acid from the sulphides. Treatment of the muds with sodium azide (a respiration inhibitor), prevented the development of acidity, indicating that the acid was produced biologically. Further analysis showed that the increasing sulphate content of the soil was accompanied by an increase in the numbers of *Thiobacillus* cells, becoming active as the muds became aerobic. Although these bacteria grew best at pH 5·0, they caused the soil pH to drop to 0·6 but below pH 3·0 the numbers of *Thiobacillus* cells declined because of an increase in the amount of dissolved toxic ferric and aluminium ions. Hart found that liming prevented this decrease in cell numbers and increased the amount of sulphur oxidation, but addition of more than 8 mg lime per 100 grams soil led to further inhibition as the pH rose above 6·0. In soils with a population of several different species of *Thiobacillus*, the effect of pH is not so clear-cut for the combined optimum ranges for *T. thiooxidans*, *T. intermedius*, *T. thioparus* and *T. novellus* cover values from 2·0 to 9·0. Thus knowledge of the species composition of soils is important if reclamation is to be attempted. Also as Hart[192] has pointed out, the rate of oxidation of polysulphides is governed by their particle size, small particles being attacked more rapidly, so that these factors may affect reclamation as well. The sulphate formed during these oxidation processes combines with calcium carbonate and produces gypsum. This plays an important role in converting marine sodium clays to terrestrial calcium clays. If the clay remained unmodified, the large quantities of sodium ions would be toxic to plant growth, but when they are displaced from the clays they are leached out of the soil during the early stages of reclamation[524].

Hydrogen Oxidation

There are a number of bacteria found in soil which possess a hydrogenase enzyme and can oxidise hydrogen to provide energy and reducing power for the fixation of carbon dioxide. These organisms are facultative autotrophs and in most instances the energy they obtain from autotrophic reactions is insignificant when compared with the energy they liberate from the decomposition of organic matter. Nevertheless, the inorganic transformations in which they participate are sometimes very important in the soil; in addition to oxidising hydrogen, some of them reduce sulphate or cause denitrification.

(a) *Sulphate-reducing Bacteria*

Sulphate-reducing bacteria can use molecular hydrogen to reduce sulphate and provide energy for carbon dioxide fixation[373].

$$SO_4^{--} + 4H_2 \xrightarrow{\text{sulphate reductase}} S^{--} + 4H_2O$$

Stüven[496] has shown that the cells can also incorporate acetate during sulphate respiration as well as carbon dioxide, and Mechalas and Rittenberg[316] reported that organic compounds were utilised as energy sources, e.g. yeast extract.

The best known sulphate-reducing bacterium is *Desulphovibrio desulphuricans*, a Gram-negative, rod-shaped bacterium, but there are also some spore-forming, sulphate-reducers now placed in the genus *Desulfotomaculum*. Adams and Postgate[2] found that *Desulfotomaculum* species were more widespread in soil than *Desulphovibrio* species; the latter were more characteristic of aquatic environments.

Postgate[374] has reviewed the ways in which sulphate-reducing bacteria can alter their environment: the effects include formation of hydrogen sulphide and displacement of oxygen, generation of an alkaline pH through the hydrolysis of sulphide ions, removal of heavy metals, e.g. iron, from solution, removal of hydrogen (a frequent product of anaerobic metabolism) and sulphate, elimination of aerobic organisms and methane bacteria and stimulation of blue-green algae and thiobacilli. Fortunately, sulphate-reducing bacteria do not dominate soils except under waterlogged conditions, when they cause toxicity as in rice paddy soils or participate in the phenomenon of gleying in the lower horizons of soil profiles. In addition, they can dissolve soil phosphates[465] and corrode iron pipes buried in the soil by removing hydrogen ions from the iron surface, causing electrolytic dissolution[474]. Their development is favoured by organic matter which may even be a prerequisite for hydrogen sulphide formation in certain soils[509].

(b) Other Hydrogen-oxidising Bacteria

Other hydrogen-oxidising bacteria can use both oxygen and nitrate for hydrogen acceptors, and in the latter case carbon dioxide fixation is accompanied by denitrification. An organism of this type is *Micrococcus denitrificans*[528]. When nitrate is utilised, growth is very slow and it is clear that this organism obtains most of its energy and carbon from organic compounds. Nothing is known of the ecology of this bacterium, but one may conclude from its behaviour in culture that it plays little part in denitrification in the soil.

The true hydrogen-oxidising bacteria, or Knallgasbacteria, obtain energy for the assimilation of carbon dioxide or organic compounds from the oxidation of hydrogen with oxygen. These aerobic bacteria include species of *Hydrogenomonas*, *Nocardia* and *Streptomyces*. *Hydrogenomonas* species are widespread in soil[430] but their significance is obscure since freshly isolated strains are physiologically variable and difficult to keep under laboratory conditions.

Many other organisms possess hydrogenase enzymes, although not all of them are able to fix carbon dioxide. *Azotobacter* has a hydrogenase which is used for the reduction and fixation of atmospheric nitrogen.

Other Autotrophic Bacteria

Ferrous iron is uncommon in most soils but it may be converted to ferric iron to yield energy by the iron bacteria, *e.g. Gallionella*. Pringsheim[381] has reported that ferrous iron is often found where water containing humic acids flows out of swamps into calcareous soils, and iron bacteria are common in such places. However, ferrous iron may be converted to ferric iron spontaneously in alkaline soils and by a variety of heterotrophic bacteria.

No significant information is available on the occurrence and distribution of other autotrophs, e.g. carbon monoxide and manganese oxidising bacteria.

Carbon Fixation by Algae

Unlike the other microbial autotrophs, algae may be significant in soil with regard to carbon fixation. In most soils, the major source of carbon is the higher plant, but there are a number of environments in which higher plants do not grow or in which they grow poorly, e.g. rock surfaces, lava flows, deserts, salt marshes and young sand dunes, and in these environments algae are often abundant.

Jacks[223] has suggested that as rocks are weathered, they become less suited to colonisation by micro-organisms since small particles are produced which may become compacted, leaving little room for air or water. Nutrients are rapidly leached away by rainwater and the total effect is one of progressive loss of chemical and physical energy. This process is reversed when rocks are colonised by algae and lichens, which not only supply the initial organic matter in the newly forming soil, but also produce metabolites which hasten weathering and further release of nutrients. It has been claimed that both green and blue-green algae and diatoms are able to change rocks into amorphous silica, montmorillonite and beidellite[371]. Lichens are active in attacking limestone, causing the accumulation of exchangeable calcium and the removal of phosphorus and sulphur from the rocks. Since they secrete chelating agents, they are able to alter the availability of iron and aluminium in the soil. Krasilnikov[256] has shown that autotrophic bacteria also occur on rocks and it is known that thiobacilli can dissolve rock materials. Whether microbial colonisation was the first stage in the formation of most soils is controversial. Unless climatic changes occurred, it is likely that the algal/lichen associations would have remained the only colonisers; higher plants would have been unable to grow because of insufficient rainfall or extreme temperatures. The colonisation of lava flows by algae is followed by the growth of other plants.

The types of algae which colonise barren surfaces are determined by the climate. *Gloeocapsa*, *Gloeotheca*, *Aphanocapsa*, and *Nostoc* (one of the more efficient silicate decomposers) occur in humid regions whilst *Protosiphon*, *Anacystis* and others occur in dry climates[444]. In temperate regions, green algae are more prominent, e.g. *Palmogloea*, although blue-green algae such as *Calothrix* are common in salt marshes[482].

It is generally assumed that algae and lichens provide organic matter enabling other organisms to colonise the soil. However, there have been no accurate estimates of the amount of carbon that these organisms fix, although in recent years it has been shown that many of the blue-green algae fix nitrogen (p. 136). Apart from accumulating organic matter, algae prevent erosion in arid soils and Booth[39] found that various blue-green algae formed a continuous cover on some desert soils, binding the soil particles together and protecting them from the force of falling rainwater. It has been suggested that in less arid climates, algal crusts may be important because they prevent leaching and provide a reserve supply of essential nutrients[13]. In rice paddy soils, although nitrogen fixation by blue-green algae is of great importance, Alexander[6] has suggested that the oxygen they liberate during photosynthesis enhances the growth of the submerged rice roots.

9

Interactions between Soil Micro-organisms

It is generally agreed that soil micro-organisms live in a highly competitive society, where many interactions occur between different species, which affect their growth and survival. It is wrong to assume that all soil microbes are in direct competition with each other, for interactions will occur only between those competing for the same or closely related ecological niches, and an interaction between two soil isolates observed in laboratory conditions may never occur in the soil. Within an ecological niche, competition for nutrients, space, oxygen and other essentials will be intense. As in all highly competitive societies, much exploitation occurs and some members resort to foul play, eliminating their rivals by poisoning, stabbing or strangling them and even on occasions inducing them to commit suicide! Such interactions, in which at least one of the microbes is harmed, are termed antagonistic[348] and include parasitism, antibiosis and predation. Not all members of the soil population are so callous and some are charitable enough to provide nutrients or growth factors for their less fortunate colleagues! These interactions, whether they are antagonistic or not, vary considerably in their specificity. While a parasitic or symbiotic relationship may be highly specific, the effects of antibiotics and growth factors on soil populations may be quite random.

Specific Interactions

(*a*) *Parasites of Bacteria*

Recent studies by Stolp and Starr[485] have shown that small, vibrio-like bacteria, which they called *Bdellovibrio*, could attach themselves to Gram-negative bacterial cells and cause lysis. *Bdellovibrio* cells are motile and when they hit a host, they stick to it. It is now known that they attach themselves to the host with the aid of a projection from one end of the cell, and in some instances a swollen hold-fast is also formed (Plate IX)[445]. Whether they dissolve their way through the wall or actually stab the host is not certain, but extracellular proteinases extracted from the parasite will not damage intact host cells. *Bdellovibrio* strains appear to be widespread in soils and aquatic environments, although only 40 to 50 cells per gram have been reported from soils. The *Bdellovibrio* strains studied so far appear to have a clear host specificity.

More familiar parasites of bacteria are phages which may cause lysis of host cells or form a stable association with the bacterium. The latter are called temperate phage and the hosts referred to as lysogenic, since the virus may at any time revert to a destructive parasitic existence. While some phages are specific to a particular strain of bacteria, others are polyvalent and attack bacteria from different genera. Although phages can be isolated readily from most soils, it is unlikely that they cause large-scale destruction of bacteria in soil. Crosse[99] found that soils enriched with concentrated suspensions of plant pathogenic *Pseudomonas* species rarely yielded more than 100 phage particles per ml of soil suspension, this being commensurate with the lysis of only two or three cells. Phages attacking strains of *Rhizobium japonicum* have been detected in both soil and in root nodules[255]. In the soil, up to 1,000 phage particles per gram were found, whilst in the nodules there were 100 times as many. Most of the phages isolated were highly specific and attacked only the strain from which they were isolated. The increase in numbers in the nodules is surprising since it is generally assumed that very few rhizobia penetrate the nodule, the bacteroids arising from only a small number of cells. Possibly many of the rhizobia that do penetrate nodules are lysogenic and the change in environment causes induction and liberation of phage. Alternatively the differences may be an artefact caused by the different density of soil and root nodule tissue.

Viruses which attack fungi (including basidiomycetes and yeasts) and blue-green algae have also been reported, but nothing is known of their distribution in nature.

(*b*) *Symbiosis with Soil Animals*

There are many micro-organisms which live symbiotically with animals that have larval stages in the soil (e.g. *Coleoptera*, *Diptera* and *Hymenoptera*). These partnerships, at least the permanent ones, are usually formed with animals inhabiting the humus rather than less decomposed litter on the surface of soil. Possibly there is little readily available food for the animals in the humus and it has been suggested that the symbiotic microbes may synthesise essential nutrients lacking in this region of the soil[53]. Thus, when yeasts are eliminated from the gut or gut cells of certain beetles, the animals grow poorly unless B vitamins and yeast sterols are added to their diet; some gut bacteria have the capacity to digest cellulose and chitin and the breakdown products may be used by the animal.

The presence in the animal gut of micro-organisms which possess a greater enzyme potential than the host may be quite common, but the small size of many soil animals has made experimentation very difficult. The microbes involved in these relationships show different degrees of integration with their hosts. Some occur intracellularly in cells around the gut, while others occur in the gut lumen; the former appear to be confined to the animal and do not have an active free-living phase, while the latter include soil microbes passing through the gut. Transmission of the micro-organism from one animal to another in the more specific partnerships occurs through infection of the eggs, either before or after laying; in the latter case, eggs may be infected from faecal material. Some beetles and crane flies have larvae that hatch in the soil and which are apparently infected by chance; by the time the adult stage is reached, they have an almost pure culture in their guts and other soil micro-organisms have been eliminated.

A looser relationship is exhibited by animals which cultivate microbes and then eat them. The wood-boring ambrosia beetles grow a variety of fungi in their tunnels, such as *Ceratocystis*, *Cladosporium*, *Endomycopsis* and *Penicillium*. The growth of these fungi, often associated with blue staining of timber, may be confined to wood but they also occur in soil. Ants also cultivate fungi, often as pure cultures on faecal pellets and leaves[17]. The maintenance of these pure cultures has been studied recently by Martin *et al.*[579] who found that ants deposit liquid on these leaves. This liquid has no antibiotic activity but contains enzymes which attack protein in the leaves and allow the cultivated fungus to grow on the breakdown products. The cultures remain pure because the ants scrape the leaves free from other fungi and spores, shred the leaves into pieces and treat each fragment with a massive inoculum of the required fungus. The maintenance of the system thus depends upon the careful preparation of a favourable micro-environment rather than the selective suppression of unwanted organisms.

(c) *Predacious Fungi*

Nematode worms, which are widespread in soil and often cause serious damage to crop plants, may be attacked by a variety of soil organisms, including the predacious fungi. These fungi have structural adaptations to trap and penetrate nematodes, including the production of sticky mycelium and spores and constricting loops which strangle the worms. Some of the fungi are internal obligate parasites which digest the nematode and then produce sticky spores on a few hyphae which grow out from the corpse. These spores stick to healthy nematodes, germinate and then penetrate them. The most common fungi of this type are *Acrostalagmus* and *Harposporium.*

The most fascinating predacious fungi are the external facultative parasites, for it is these forms that have evolved the most elaborate trapping mechanisms. Some species of *Arthrobotrys* produce a network of sticky mycelium in which the nematodes become trapped (Plate IX). Others form adhesive stalked knobs (e.g. *Dactylella ellipsospora*) or have non-adhesive stalked rings (e.g. *Dactylaria candida*). In the latter forms, nematodes push their head or tail into the rings and become wedged. Even more diabolical are the fungi which produce constricting ring traps, such as *Dactylella bembicoides.* The rings are formed from three cells and project on a short stalk. A few seconds after the nematode touches the inside of one of these rings, the cells swell and crush the unfortunate worm. The swelling occurs in only one-tenth of a second, suggesting that some physical change occurs in the cells, possibly in the wall or membrane structure. However the worms are caught, they die and then absorbing hyphae grow into them and digest the internal contents of the body. Not surprisingly, the murderous habits of these fungi are fascinating to watch and can glue an observer to the microscope as readily as nematodes to their mycelium!

In the absence of nematodes, most of these fungi grow well and sporulate on agar media. When nematodes, or filter-sterilised water that has contained nematodes, are added to such cultures, trap formation is stimulated. These solutions have been found to contain a water-soluble protein, termed nemin, different concentrations of which are required to stimulate trap formation in different fungi[259, 378].

Despite the intensive studies of the morphology and physiology of these fungi, little is known about their growth in soil. Surveys by Duddington[119] and others have shown that predacious fungi are commonest in damp habitats, such as leaf litter and moss cushions, although they occurred in all soils studied, with the exception of a few mineral soils and acidic peats. The commonest species in English arable soils were *Arthrobotrys oligospora*, *Harposporium anguillulae*, *Acrostalagmus obvatus* and an unidentified member of the *Zoopagales*[123]. Data on their abundance in soil are scanty, since most mycologists do not encourage the growth of nematodes on isolation plates. Furthermore, rapidly growing saprophytes overgrow and obscure these fungi, which do not appear on isolation plates until two or three weeks have elapsed.

Interest in the behaviour of these fungi does not stem entirely from morbid curiosity, for it has been suggested that they might help to control the incidence of root-infecting nematodes in soil. Linford[271] showed that green manuring of soils increased the numbers of both nematodes and predacious fungi. Cooke[93] indicated that the increase of predacious fungi was not due to the increase of nematodes, but to a direct effect of the added organic matter on the fungi. The effect of organic additions is only temporary, however, and predacious activity soon declines. Furthermore, additions of sucrose to soil led to an increase in nematodes but a decrease of trapping. The problem of tipping the delicate equilibrium between host and parasite in favour of the parasite is obviously difficult, and efficient biological control has not yet been achieved.

Like nematodes, protozoa are attacked by a variety of micro-organisms in soil, including predacious fungi. Most of these, which may be either internal or external obligate parasites, are in the order *Zoopagales*. Infection of protozoa, usually amoebae but occasionally rhizopods, is achieved by growth of the fungus from spores or mycelium which stick to the animals or are ingested by them. A well-known example of such a fungus is *Cochlonema verrucosum*. There is great host specificity amongst these fungi, at any rate in the laboratory, for they are usually confined to only one species of protozoa.

Arthropods are parasitised by many microbes, such as the fungus *Cordyceps* which attacks the larvae of various insects and spiders and *Bacillus thuringiensis* which infects insects. The status of such micro-organisms in soil is far from clear.

(*d*) *Mycoparasitism*

Microbes that attack fungi range from micro-fungi attacking fruit bodies of higher fungi to viruses that infect yeasts. Many of the fungi that attack others spend only short periods in the soil. *Calcarisporium arbuscula* attacks the fruit bodies of various Basidiomycetes, and as it grows it produces a substance called calcarin which prevents the growth of saprophytic microbes on the dying remains (see Chapter 4). When the fruit bodies have disappeared, *Calcarisporium* forms sclerotia which help it to survive in the soil from one fruiting season to the next[552]. It may also survive in association with mycorrhizal mycelium on tree roots and on fragments of chitin-containing substrates[155].

During growth in culture, *Calcarisporium* can, in turn, be parasitised by *Trichoderma viride*. This species may produce antibiotics which weaken the fungal host before infection, although many mycoparasitic fungi require an intimate association between host and parasite before infection can be initiated. Such parasites may penetrate their hosts with small pegs, produced from hyphae coiling around or growing adjacent to the host mycelium (e.g. *Penicillium vermiculatum*). Possibly such forms produce lytic enzymes which partially digest the cell wall and make penetration easier. Examples of fungi which have been observed parasitising other fungi are *Rhizoctonia solani*, *Didymella exitalis* (which can attack the wheat pathogen *Ophiobolus graminis*) and *Gliocladium roseum*[38].

Much of the evidence for the occurrence of mycoparasitism in soil rests on observations of colonisation of lysed hyphae, but it is rarely clear whether the hyphae were colonised, infected and lysed before or after their death. Furthermore, when penetration of hyphae has been observed, as in the case of *Aspergillus niger* by *Streptomyces albus*[393], it has usually been seen on glass slides rather than soil particles. This may be important as Lockwood[278], who mixed fungal mycelium with soil and recovered it later on adhesive plastic film, found that mycelium was never colonised by micro-organisms until it had been almost completely destroyed. This suggests that autolysis, followed by colonisation, is a more frequent occurrence than infection followed by heterolysis. Autolysis is usually the result of starvation, but can also be induced by lack of oxygen and the accumulation of metabolites toxic to the producing organism[47]. This makes it difficult to distinguish between autolysis and heterolysis. Other evidence for suggesting that mycoparasitism is uncommon in soil has been provided by Boosalis and Mankau[38], who found that even under the most favourable conditions, only 18 per cent of *Rhizoctonia* mycelium inoculated into soil, and about 0·06 per cent of naturally occurring hyphae, were parasitised.

Not surprisingly, fungal spores are usually more resistant to lysis than vegetative mycelium. Often lysis occurs immediately after germination[83], suggesting that a fungus will be able to survive in soil for a longer period if the environment is unfavourable for germination.

Non-specific Interactions

(a) Antibiosis

An antibiotic is an organic substance produced by a micro-organism and which, acting in very low concentrations, can kill or inhibit the growth of another micro-organism. Broader definitions have been

Table 26. *Micro-organisms producing some well-known antibiotics*

Producer	Antibiotic
Streptomyces antibioticus	Actinomycin
S. erythraeus	Erythromycin
S. fradiae	Neomycin
S. griseus	Streptomycin
S. niveus	Novobiocin
Bacillus polymyxa	Polymixin
Penicillium chrysogenum	Penicillin

made and under these many microbial metabolites may be regarded as antibiotics; Park[348] has pointed out that products such as carbon dioxide and cyanide can cause antibiosis, but here we will confine our attention to antibiotics in the more limited sense. All antibiotics are, to some extent, specific in their effects. However, some antibiotics have a relatively broad spectrum, e.g. streptomycin, which is active against many Gram-positive and Gram-negative bacteria and actinomycetes, while others have a more limited spectrum, e.g. viomycin, which is active mainly against *Mycobacterium* species.

Over the last 25 years, hundreds of antibiotics have been found in laboratory cultures of soil micro-organisms, including some fungi and bacteria, but mostly in actinomycetes of the genus *Streptomyces* (Table 26). Although the anti-microbial activities of purified antibiotics are well known, there is still some doubt about their role and importance in soil. The potential value of antibiotic production to a micro-organism in a competitive society is clear, but direct proof that antibiotics are produced in soil is lacking. Few workers would argue that a micro-organism's activities in culture are identical to those in soil and it is possible that antibiotic production could be simply an accidental occurrence caused by the highly unnatural conditions of culture. Therefore, one must look for evidence of their occurrence in the natural habitat.

Their production has never been shown in completely natural soil, but can be readily demonstrated in sterilised soil supplemented with nutrients, e.g. sugars and proteins. They have also been detected in unsupplemented sterile soil[152] and supplemented, non-sterile soil[229]. Circumstantial evidence for their production in natural soil was obtained by Rangaswami and Ethiraj[387], who followed the development of germ tubes from spores of *Helminthosporium*, buried in soil on slides, and observed malformations identical in appearance to those produced in culture by *Streptomyces* filtrates.

One might well ask why it has not been possible to detect antibiotics in natural soil. There are several reasons, in addition to the possibility that they may not be produced. Some are very unstable and are quickly degraded chemically or biologically in soil, while others become strongly adsorbed by colloidal particles and are difficult to extract, this being particularly true of water-soluble, basic and amphoteric antibiotics (e.g. streptomycin and terramycin)[463]. Pramer and Starkey[377] showed that streptomycin, when adsorbed, was not irreversibly inactivated and could, therefore, have a localised antimicrobial effect.

The assessment of the possible importance of antibiotics in soil largely depends on whether the system is considered on a macro- or micro-scale. Production of antibiotics will not occur in the whole of the soil mass, but will be confined to local sites where nutrients in sufficient amounts are available. Wright[575] was able to detect gliotoxin produced by *Trichoderma viride* inoculated into wheat straw and buried in soil. If it is accepted that antibiotics are only produced in localised sites for a short period of time, then it is easy to understand why they are difficult to detect in bulk soil samples where they would be present in very low concentrations, and beyond the sensitivity of analytical methods. The sensitivity of a direct assay method and a bioassay technique was compared by Soulides[462]. Antibiotic-clay complexes, which contained amounts of antibiotics in excess of those adsorbed by the clay, were made; the antibiotics were then extracted and amounts recovered assayed. The bioassay technique was found to be highly sensitive for the detection of certain antibiotics and minimum detectable amounts are given in Table 27.

Table 27. *Minimum detectable quantities of antibiotics in soil determined by a direct assay method and an extraction-bioassay method*[462]

Antibiotic	Direct assay method	Extraction-bioassay method
	μg of antibiotic per g of soil	
Streptomycin	60·00	5·00
Carbomycin	15·00	0·30
Terramycin	5·00	0·08
Aureomycin	3·00	0·02
Bacitracin*	15·00	0·15

* Figures refer to units of bacitracin.

One might expect that antibiotic-producers, and those resistant to them, would be at a great advantage in competitive interactions in soil and would therefore become dominant and numerous. However, this is not the case. It must be remembered that production or tolerance of antibiotics would be only two of the many factors which determine the outcome of competition in soil, and as already emphasised, all soil microbes are not in competition with each other. Only in the case of those competing for the same substrate, in the same place and at the same time, will they have any influence on the interactions. Unfortunately, there are no techniques available which are sufficiently precise to pin-point the microbes involved in such subtle interrelationships.

Another ingenious argument for antibiotic production in soil has been provided recently by Pollock[370]. Some bacteria (e.g. *Bacillus licheniformis*) which are potentially sensitive to penicillin can produce an enzyme called penicillinase, the sole function of which is to break down this antibiotic. If antibiotics are entirely unnatural products, one would expect penicillinase production to be a recently acquired capability, as penicillin has been in use only for the last 30 years. However, it was found that *Bacillus* strains, resuscitated from endospores present in dried soil on plants which had been in the British Museum since 1689, also produced penicillinase.

(*b*) *Fungistasis* (*Mycostasis*)

While much of the evidence for the occurrence of antibiotics in soil is circumstantial, most soil microbiologists find it difficult to believe that they are not important. As we have seen, it is likely that they act for limited periods in localised sites, but there are also examples of large-scale and long-term inhibitory effects in soil which must be explained. Dobbs and Hinson[115] showed that the germination of many fungal spores could be prevented by an unknown factor in soil. The effect was detected when viable spores of certain common soil fungi, e.g. *Mucor ramannianus* and *Penicillium frequentans*, were placed in direct contact with soil or on surfaces such as cellophane film placed on or in the soil. Similar effects have been observed in soils throughout the world, including Great Britain[115], America[272], Africa[224], and Malaya[173]. Wherever this phenomenon occurs, it has the following characteristics:

(*i*) While the germination of many fungi is inhibited, some are unaffected;

(*ii*) The inhibition occurs in the presence of microbial activity, but can be removed by the addition of nutrients, e.g. glucose, or sterilisation of soil, and re-introduced by addition of non-sterile soil or soil microbes;

(*iii*) The inhibitory principle is water soluble and diffusible.

All this suggests that the static effect is of microbial origin. Lockwood and Lingappa[279] tested the effects of many soil microbes on the germination of spores of the fungus *Glomerella cingulata* in sterile soil. The results obtained (Table 28) show clearly that many of the isolates

Table 28. *Frequency with which isolates of actinomycetes, bacteria and fungi inhibited germination of conidia of* Glomerella cingulata[279]

Isolates	Proportions of cultures causing inhibition of germination	
	On agar medium	In soil culture
Actinomycetes	11/20	20/20
Bacteria	14/27	27/27
Fungi	8/20	20/20

inhibited spore germination. The possibility that antibiotics are involved immediately springs to mind, but if this is so, it is surprising that addition of nutrients to soil removes the effect. Also, it is unlikely that antibiotics could persist long enough or be so widely distributed in the soil mass. Lockwood[277, 278] suggested that microbial competition for limited supplies of exogenous nutrients was a more likely explanation. In the presence of living fungal spores or their exudates, numbers of bacteria and actinomycetes increased in the soil[272], suggesting that nutrients necessary for spore germination might be removed by microbial activity. Ko and Lockwood[253] have provided other evidence for the importance of competition for nutrients in fungistasis, by studying the behaviour of fungal spores in soil and in sterile distilled water. They found that fungi, whose spores germinated in distilled water and therefore did not need exogenous nutrients, were insensitive to fungistasis, e.g. *Neurospora tetrasperma*. However, spores of other fungi, e.g. *Glomerella cingulata*, which were sensitive to fungistasis, germinated in distilled water in static conditions but not if they were leached with running water. It was suggested that leaching was analogous to removal of nutrients from the vicinity of the spore by microbial activity in soil. Finally, there were those spores which depended on the presence of exogenous nutrients for germination but were unable to germinate in distilled water and were sensitive to fungistasis, e.g. *Mucor ramannianus*. Spores of this latter group could be induced to germinate in soil only when nutrients, such as alfalfa residues, were added. These results and interpretations are fully consistent with the previously outlined characteristics of fungistasis.

Other workers believe that the static factor is essentially a product of the soil rather than micro-organisms, while some suggest that it is caused by toxic breakdown products of substances such as lignin. Dobbs and Gash[114] attempted to reconcile these results by suggesting that there might be two types of fungistasis, one of microbial origin (removed by sterilisation or addition of nutrients) and the other a residual fungistasis, which does not react to these treatments.

It is interesting to speculate on the importance of fungistasis to the growth and behaviour of fungi in soil. It is clear that some fungi (e.g. the sugar fungi) exist in soil as spores for much of the time and have only short periods of active growth in sites where suitable nutrients allow the germination of their spores (see Chapter 4). These characteristics are also found in many root-surface fungi, whose spores do not germinate in the absence of nutrients from root exudates or dead cells; Jackson[225] showed that seedling roots could break the fungistatic effect in soil.

A similar widespread bacteriostatic effect has not been observed, but occasional reports of inhibitory factors in certain soils have been made. Thus the germination of *Azotobacter* cysts is inhibited in soil and the effect can be removed in the presence of pea roots, presumably by their exudates[55].

(c) *Growth Promotion Effects*

The succession of micro-organisms on a substrate in soil is often brought about by one wave of colonisers making nutrients available for the next, and this is probably the most widespread kind of growth promotion effect in soil populations (see Chapter 4). However, in addition to this, many microbes can synthesise vitamins, at least in culture. In some cases, leakage of growth factors into the surrounding medium occurs, which may stimulate the growth of other microbes unable to synthesise their own vitamins. Lochhead and Thexton[276] reported that filtrates of cultures of soil bacteria were capable of promoting growth of other soil bacteria. Cook and Lochhead[92] found that the capacity for synthesising growth factors in culture was most marked in bacteria isolated from rhizosphere soil; while 37·4 per cent of soil isolates produced one or more growth factors, 79·8 per cent and 79·0 per cent respectively of rhizosphere and rhizoplane isolates could do so. The growth factors most frequently produced by the root region bacteria were riboflavin and nicotinic acid, although synthesis of a wide variety of growth factors was demonstrated by Lochhead[275] and Pantos[344]. A clear interaction between two bacteria was shown by Lochhead and Burton[274], who discovered a hitherto unknown vitamin, the *terregens* factor, produced by *Arthrobacter terregens* and required by *A. pascens*.

It is also known that several fungi and actinomycetes isolated from soil produce growth factors in culture; within the genera *Penicillium* and *Aspergillus* there are a number of vitamin-producing species[90] and Georgieva and Sheikova[146] reported the synthesis of vitamin B_{12} by various actinomycetes from Bulgarian soils.

Although it is known that vitamins are present in soil, little is known of their origins. Lilly and Leonian[270] detected thiamin and biotin in soil extracts and Roulet[409] found 1·9 μg of thiamin and 0·06 μg of biotin per 100 g of surface soil. These, however, were not necessarily produced by micro-organisms, for they could have originated from plant litter or roots. Concentrations of riboflavin over 500 μg per g have been detected in forest litter[436] and enough biotin to promote microbial growth is produced by plant roots[418]. Only small quantities are needed to stimulate microbial development in culture and Meshkov[321] showed that while 0·01 μg per ml of thiamin promoted growth of *Pseudomonas aurantiaca*, 0·02 μg per ml decreased growth; *Bacillus mycoides* was inhibited at 0·01 μg per ml but *Azotobacter* was stimulated by concentrations of biotin as low as 0·002 μg per ml. There is evidence that a requirement for growth factors is to some extent related to the location of the microbe in soil. Cook and Lochhead[92] showed that micro-organisms isolated from the root region of wheat (*Triticum*) had less requirement for growth factors than did those from root-free soil.

Although the production and effects of growth factors in culture conditions have been clearly demonstrated, their importance in soil remains to be assessed. Probably the most convincing evidence for vitamin production by microbes in soil was provided by Schmidt and Starkey[436], who added plant materials containing small amounts of vitamins to soil. Initially, action of soil microbes on the residues resulted in an increase in vitamins, the peak amounts coinciding with peak carbon dioxide evolution, i.e. maximum microbial activity. This provided indirect evidence for the microbial synthesis of vitamins in soil sites where nutrient concentrations were high.

10

Micro-organisms in the Soil-plant Ecosystem

This book has been largely devoted to a consideration of the role of micro-organisms within micro-environments and ecological niches of a heterogeneous soil system. To a large degree, this reflects a change in direction in research over the last twenty years, away from the consideration of the soil as a single habitat and the idea that all processes in it are integrated into neat cyclic patterns. However, it is essential that the overall importance of the soil microflora is not forgotten, so in this final chapter we shall consider the activities of the soil micro-organisms discussed earlier in a more general context, relating them to the activities of other organisms living in or on the soil.

The Ecosystem Concept

The importance of the soil microflora may be evaluated by determining the amounts of energy and matter which are diverted and used by it. For this purpose, it is convenient to regard the soil as an ecosystem, that is an open system in a steady state in which the quantity of energy entering is exactly equalled by the amount of energy ultimately surrendered by it[364, 52]. In this sense, the soil may be compared with an individual organism or organ and it is interesting to note that as early as 1946, Quastel[384] had suggested that the soil was analogous to the liver in terms of energy flow.

It follows, if the soil is viewed as an ecosystem, that 'production and consumption of each element of the system are exactly balanced, and the concentration of all elements remains constant'[52]. This might seem a strange description of the soil in view of the importance we have attached to its heterogeneity and the day-to-day and point-to-point fluctuations that occur in the microbial populations within it.

However, these fluctuations and variations occur on a small scale and if the soil is looked at over a long enough period and on a large enough scale, approximately steady state conditions obtain and have been referred to by microbiologists as a dynamic equilibrium[58]. Thus population size is more uniform when measured on an annual basis rather than on a daily basis. Similarly, a comparison of population sizes between a root fragment and the surrounding soil may well show differences, while a comparison of two large, representative samples taken from soil in which roots are growing may reveal negligible differences because the total sum of small-scale variations is the same in both samples.

If a progressive change in the population and environment occurs over a long period of time, we must conclude that the system has not matured and that succession is taking place. Garrett[142] has pointed out that successions of heterotrophic organisms progressively deplete their substrates and simplify the environment (see also p. 78). If the substrates are exhaustible, then the end-point of the succession is not a complex, structured climax, but is zero. However, if the substrates are being supplied continuously to the soil, e.g. through litter fall and root growth, then ultimately a balance between input and output is set up and a stable ecosystem results. The point of equilibrium is determined by the interactions between the organisms and their environment. Since the routes by which this equilibrium is reached are varied, Brock[52] has suggested that studies of succession may be thought of as studies of the 'embryology of ecosystems'.

Food Chains within the Ecosystem

In most terrestrial ecosystems, the major energy input is from the sun. This solar energy is converted into chemical energy by plants through the process of photosynthesis and is used by them for various purposes including growth and respiration. During respiration some energy is dissipated as heat, a form of energy which is unavailable to other living organisms and which is therefore lost from the ecosystem. Some of the chemical energy stored in plant material is removed by grazing, but the rest ultimately reaches the soil in the form of dead leaves, detached woody structures and living roots. These tissues are then attacked by the decomposers, which comprise herbivorous soil animals and the soil microflora. In many ecosystem analyses, it is useful to recognise the existence of two pathways of energy flow, the 'grazing food chain' and the 'detritus food chain'. In the grazing food chain, living organisms are successively eaten by other organisms, e.g. phytoplankton⟶*Daphnia*⟶fish⟶man; while in the detritus food chain, dead organic material is consumed, e.g. dead leaves⟶earthworms (which die) ⟶fungi (which die)⟶bacteria. In the soil it is virtually impossible to distinguish between these two pathways. Many soil animals, which are part of the grazing food chain, feed on

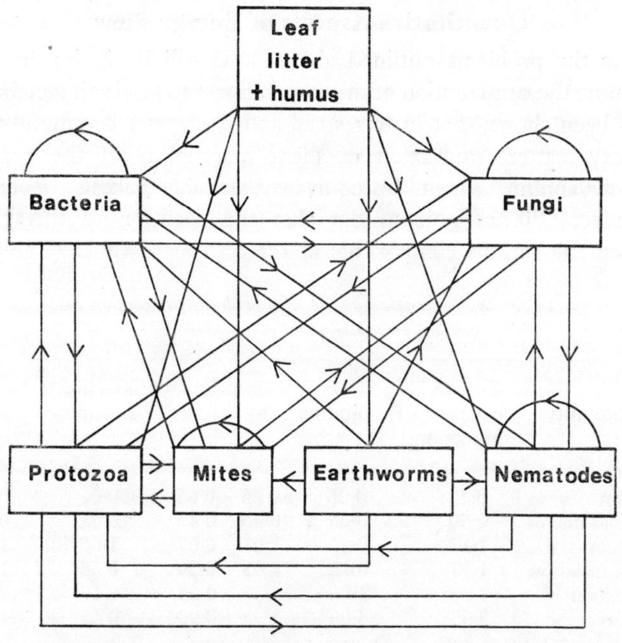

Fig. 24. *Food web showing possible interrelationships between soil organisms and their substrates*

micro-organisms in the detritus food chain attacking dead leaves; they also provide substrates in the detritus food chain in the form of corpses and faecal material. The two chains are hopelessly and inextricably tangled and form an exceedingly complicated food web (Fig. 24).

In analyses of plant-animal food chains, it is often possible to discern a definite sequence of events, e.g. A eats B eats C eats D, but in food chains involving micro-organisms this is rarely possible. Micro-organisms are more indiscriminate in their choice of substrates, partly because of their greater enzymic potential and partly because they can attack substrates of many different sizes. Thus fungus P may attack any of the substrates X, Y and Z in any order and at any time, e.g. it could attack freshly fallen leaf litter, material sloughed off animals that feed on this leaf litter or even the dead remains of these animals. It will be readily appreciated, therefore, that simple models of these systems cannot be constructed and as a result microbiologists have concentrated more on the problems of succession in micro-environments rather than on food chains in the ecosystem.

Quantitative Aspects of Energy Flow

Despite the problems outlined above, it is still intriguing to try to determine the contribution of micro-organisms to total soil metabolism. It has been shown that in terrestrial and freshwater communities the nett dry matter production of plant material is of the order of 1 kg/m^2/annum which is equivalent to 4,800 kcals of energy. Macfadyen[289] has pointed out that the partition of this energy between the various components of the ecosystem varies (Table 29).

Table 29. *Partition of the nett primary production in various ecosystems (kg/m^2/annum)*[289]

Community	Gross* primary production	Respiration	Nett primary production** Total	Eaten	Decomposed	Mean*** stock
Plankton	0·72	0·06	0·65+	0·65	<0·01	0·004
Algae-salt marsh	0·50	0·05	0·45+	0·45	<0·01	0·003
Spartina marsh	1·17	0·10	1·07	0·07	1·00	1·06
Grazed meadow	1·17	0·12	1·05	0·39	0·66	1·00
Beech wood	2·35	1·00	1·35	0·95	0·40	15·5
Rain forest	5·35	4·00	1·35	0·90	0·45	24·0
DERIVED PERCENTAGES						
Plankton	100	9	90+	90	<5	0·6
Algae-salt marsh	100	10	90+	90	<5	0·6
Spartina marsh	100	9	91	6	85	90
Grazed meadow	100	10	90	33	56	85
Beech forest	100	43	57	40	17	6,600
Rain forest	100	73	26	17	9	4,570

* Gross primary production—total amount of chemical energy stored by plants or other autotrophs per unit area per unit time (usually obtained by photosynthesis).

** Nett primary production—that part of the gross primary production which is available to heterotrophs, i.e. gross primary production less the energy lost in respiration.

*** Mean stock—energy content of the standing crop averaged over a period of time, usually one year.

It will be seen that the importance of the decomposers (microflora+microfauna) is greatest in the grassland communities, slightly less important in forest sites and negligible in aquatic environments. This means that comparatively little energy is available to cattle grazing on grassland and Macfadyen has estimated that only about one-seventh of the total primary production is consumed by cattle, the rest being used by other herbivores or passed through the detritus food chain[287]. When these other herbivores die, more energy passes into the detritus food chain. Plainly, the study of decomposition is very important since most of the energy stored during the primary production of plant material is released during this process. Even in the aquatic environment where nett primary production passes mainly through the grazing food chain, the study of decomposition is still important since the energy in the dead herbivores and carnivores must be liberated.

(*a*) *Measurement of Total Soil Metabolism*

The activities of soil organisms in decomposition may be assessed by determining 'total soil metabolism', i.e. the sum total of the metabolic activities of all the soil inhabitants. Usually this has been estimated rather than measured by subtracting the energy consumed by herbivores from the nett primary production. This assumes that all the remaining energy passes into the soil (and out again since the system is balanced). The measurement of energy input has been made by determining the rate of addition of organic matter to soil through leaf and twig fall. Other sources of energy have usually been ignored because of an inability to measure them, e.g. addition of faeces, movement of animals in and out of the soil, input from chemosynthetic autotrophs and synthesis of root tissue. No accurate figures of root production are available, although Macfadyen concluded from a survey of the literature that root production values lie between 15 and 50 per cent of the values for all the aerial parts of the plant[289]. Consequently estimates of total soil metabolism arrived at from these measurements of energy input will be inaccurate.

Direct measurement of metabolic activity is also possible by calorimetry (determination of heat energy evolved from the total combustion of matter) and by measurement of carbon dioxide output and oxygen uptake of soil. Although such measurements are easily obtained on laboratory samples, accurate measurement under field conditions is impossible. Thus values of soil metabolism obtained from field measurements of respiration are often in excess of those estimated by the difference and litter fall methods already referred to, due mainly to the effects of disturbance on the normal rate of metabolism and to the contribution of root respiration in the field (see also p. 65).

(b) Partition of Energy between Different Decomposers

If the problems of measuring the rate of energy input are great, those of estimating the contribution of the different microbial and animal groups are apparently insurmountable. We pass from the realm of speculation to the land of dreams!

The energy budget for any group of organisms is often expressed in terms of the following general equation:

Amount of food (energy) consumed	=	Amount of growth		Respiration losses		Excretion losses
C		G	+	R	+	E

In special circumstances, this equation may be simplified. In the case of adult soil animals, it is generally assumed that most of the energy consumed is lost through respiration, so that the sum of respiration of an individual times standing crop is a good measure of the annual energy flow through the population.

Amount of food (energy) consumed	=	Respiration of an individual		Standing crop		Respiration of whole population
C		r	×	S	=	R

Such simplified equations cannot be applied to microbial populations because micro-organisms reproduce rapidly and incorporate a good deal of the food and energy in new protoplasm[198]. Consequently, it is necessary to measure G, R and E in the natural environment if we wish to determine C.

(*i*) *Measurement of Growth* (*G*) In Chapter 3 we discussed the methods and problems of determining the biomass of micro-organisms in the soil and pointed out that measurement of growth rates in natural conditions had not been achieved. The growth rates of many soil bacteria have been calculated in the laboratory where they have been grown in nutrient-rich conditions, but the extrapolation of these results to field conditions, where nutrients are sparse, is not a credible procedure (see p. 27). The main hope for measuring growth rates in the soil is the application of tracer techniques using radioactive substances or fluorescent stains but these methods have not been sufficiently explored to prove their value.

(*ii*) *Measurement of Respiration* (*R*) The problem of measuring respiration rates of specific groups of soil organisms *in situ* has not been solved. Addition to soil of selective chemicals which inhibit all but one group has been suggested, e.g. the addition of insecticides and selected antibiotics, leaving only the fungi metabolising, but whilst this is possible it is no real solution. Many of these chemicals are insufficiently specific in action, are easily inactivated by soil and may be metabolised by the remaining micro-organisms. Furthermore, removal of competitors results in an unnaturally large amount of energy becoming available to the remaining organisms which will then thrive and grow more quickly.

Zoologists have approached this problem differently by measuring, in the laboratory, the respiration rates of animals at different stages of development[340]. Soil respiration rates have been predicted from the values obtained and from information on the size of natural soil populations. These methods are not applicable to micro-organisms as they cannot be extracted from soil in large quantities without first growing them in culture where overabundance of nutrients results in unnatural rates of respiration.

(*iii*) *Measurement of Excretion* (*E*) The concept of microbial excretion is meaningless and cannot be measured or estimated in the way that it can for animal populations. Animals live by ingesting material and passing out waste in relatively definable forms. Microbes 'excrete' enzymes into the environment and then absorb the externally digested products. In other words, the soil surrounding an organism is both its gut and its cesspool. Apart from enzymes, many other substances are secreted by the microbial cell, or just leak out. These materials are chemically diverse and yet so similar to the cell contents that it would be impossible to differentiate between them, let alone measure their rates of excretion.

(*iv*) *Measurement of C by 'Difference'* Since it is impossible to measure G, R and E with any accuracy, we cannot measure directly the energy flow through different micro-organisms. However, it has been suggested that an 'estimate' of their overall importance could be obtained by working out the total energy budget for all the other organisms and subtracting this from the total amount of energy available.

Estimates by difference are made by (*a*) subtracting the energy consumed by herbivores from the nett primary production, giving the total soil metabolism, and (*b*) subtracting estimates of the consumption of energy by soil animals from the figure so obtained. Macfadyen[286] has pointed out that these estimates are inaccurate since the often considerable errors in all the individual measurements are compounded. Nevertheless, the figures obtained in this way are of interest since they suggest that microbial metabolism constitutes about 80-90 per cent of the total soil metabolism. This does not mean that soil animals are unimportant for their absence from soil may greatly impede the action of micro-organisms on organic matter. Raw[390] showed that the elimination of earthworms led to a build-up of 0·5-1·5 inches of surface litter in apple orchards over 15 years. When earthworms were present, 90 per cent of the normal leaf fall (0·5 ton dry weight) was buried in the soil by 0·75-1·0 tons of worms, making it available for microbial decomposition. Also, as we saw in Chapter 4, the experiments of Edwards and Heath[123] emphasised the importance of earthworms, small soil arthropods, enchytraeid worms and other soil animals in initiating organic matter decay in the soil ecosystem.

Cyclic Processes in Ecosystems

We noted earlier that energy is dissipated in biological systems for it is ultimately converted into heat energy which is unavailable to living organisms. Thus energy enters the system, flows through it and is

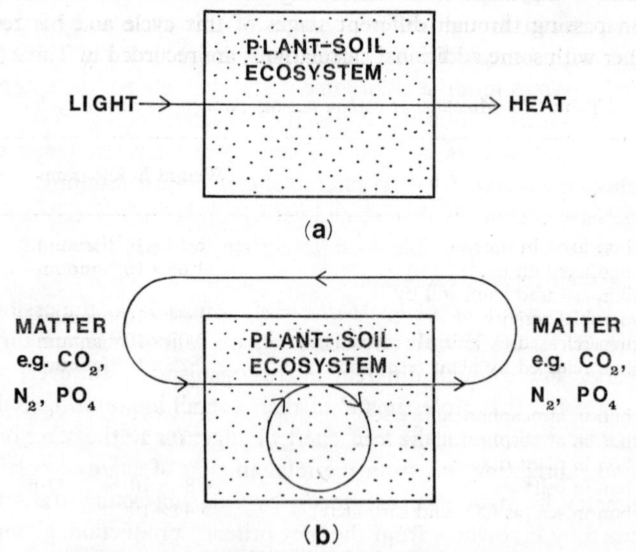

Fig. 25. (a) *Passage of energy through the plant-soil ecosystem.* (b) *Passage of matter through the plant-soil ecosystem*

then lost (Fig. 25a). This may also happen to matter that enters the system. For instance, nitrate fertilisers may be taken up by a plant and later released to be leached out of the soil. However, matter may also cycle within the system, through its conversion into a number of different chemical forms (valences) which are still available to one organism or another growing within the system (Fig. 25b). Stanier, Doudoroff and Adelberg[467] point out that any biologically important element can pass through such a cycle via a series of oxidation and reduction reactions, from the non-living environment to living matter and back again to the non-living environment, e.g. carbon, nitrogen, etc. In another form of cycling, the nutrient concerned remains unaltered in form (valence) as it passes the non-living environment to an organism's environment and back again, e.g. phosphate.

The nature of these systems may be illustrated by reference to a few examples. The individual reactions involved in these cycles have been described in earlier chapters, and the aim of this discussion is to show how these reactions link together.

(a) *The Carbon Cycle*

The carbon cycle is represented diagrammatically in Fig. 26. Loomis[280] has made some interesting calculations on the amounts of carbon passing through different stages of this cycle and his results, together with some additional calculations, are recorded in Table 30.

Table 30. *Amounts of carbon passing through the carbon cycle*

	Weight in kilograms	
Carbon fixed in the sea	$8{\cdot}17 \times 10^{13}$/annum	
Carbon fixed on land	$1{\cdot}68 \times 10^{13}$/annum	
Carbon released from soil by microbial activity	$1{\cdot}34 \times 10^{13}$/annum	
Carbon released by animal respiration	$0{\cdot}08 \times 10^{13}$/annum	
Carbon released by plant respiration	$0{\cdot}25 \times 10^{13}$/annum	
Carbon in atmosphere above sea	$3{\cdot}81 \times 10^{14}$	(0·007%)
Carbon in atmosphere above land	$1{\cdot}63 \times 10^{14}$	(0·007%)
Carbon in plant tissue	?	(40-50%)
Carbon in soil*	$4{\cdot}8 \times 10^{14}$	(1·6%)
Carbon in sea (as CO_2 and carbonate)	$5{\cdot}44 \times 10^{16}$	?

* Calculated assuming a land area of $1{\cdot}49 \times 10^{10}$ hectares and an average soil weight of $19{\cdot}94 \times 10^{5}$ kg/hectare.

It is clear from these figures that land plants fix about 10 per cent of the carbon content of the air above them and about 3 per cent of the total atmospheric supply. Oceanic plants fix a lot more, partly because of the high dissolved carbon dioxide levels in the surface layers of the sea and partly because of the greater area of the oceans.

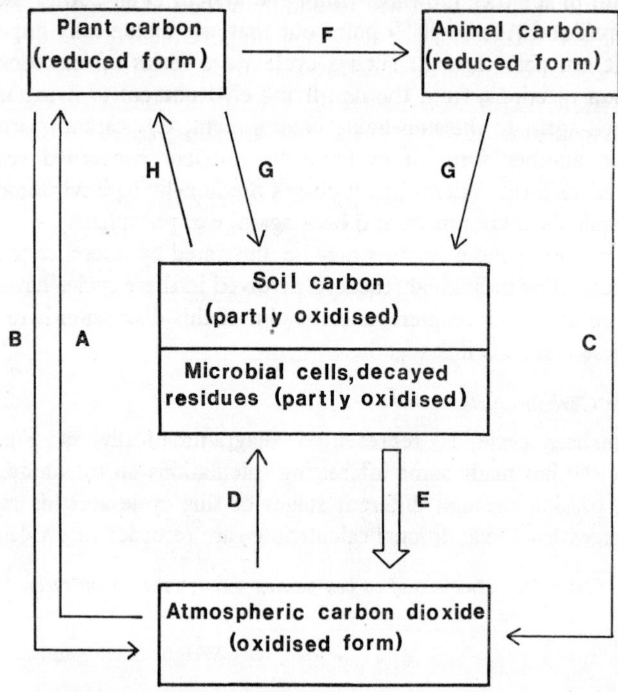

Fig. 26. *A generalised form of the carbon cycle.*
A. *Photosynthesis.* B. *Plant respiration.* C. *Animal respiration.* D. *Microbial fixation of carbon dioxide by autotrophs.* E. *Microbial respiration.* F. *Ingestion.* G. *Decomposition of dead and dying materials.* H. *Transfer of carbon by mycorrhiza*
(*Reproduced in modified form by permission of M. Alexander and John Wiley & Sons Inc.*)

The effect of carbon dioxide fixation is to concentrate carbon in plants, for whilst carbon represents about 0·007 per cent of the elements in the atmosphere (0·03 per cent carbon dioxide), it is about 40-50 per cent of the elements in plant tissue. It follows from these figures that the release of carbon from plant tissues is essential if atmospheric supplies are to be maintained and this may come about in several ways. Loomis[280] has estimated that 15 per cent of the carbon fixed by plants is lost through plant respiration, 5 per cent through animal respiration and the remainder through microbial respiration (i.e. $1{\cdot}34 \times 10^{13}$ kg/annum). The bulk of this microbial respiration will take place in the soil and since we may assume that the carbon input and output are balanced, this means that no less than $1{\cdot}34 \times 10^{13}$ tons of carbon (900 kg/hectare) will be added to the soil each year. Since much of the earth's surface does not support plant growth, greater amounts of carbon will be added to soil in the relatively fertile areas, and a recent survey of the total litter accumulation in various communities confirms this[399]; these data are summarised in Table 31.

Table 31. *Annual accumulation of total litter (leaves, above-ground perennial parts and roots) in different communities (kilograms per hectare)*

	Total litter accumulation	Carbon in litter*	% of plant biomass
Tundra—arctic	475	237·5	19
Tundra—shrub	1,235	617·5	8
Tundra—forest	2,655	1,327·5	3
Coniferous forest	1,000- 3,500	500-1,750	1·5-2·0
Deciduous forest	2,500- 3,500	1,250-1,750	2·0-4·0
Subtropical forest	9,000-10,000	4,500-5,000	?
Tropical forest	12,500	6,250	3·0-5·0
Steppeland	1,500- 5,000	750-2,500	40-50
Deserts	600- 5,000	300-2,500	30-60

* Assumes 50 per cent carbon in plant tissue.

The amount of carbon originating from above-ground parts and roots varies from plant to plant and Bray[44] found that roots comprised 13-84 per cent of the total plant in herbs and from 9 to 24 per cent in trees. It is difficult to determine when a root ceases to be part of the plant and becomes part of the soil organic matter. Consequently, estimates of the amount of carbon in the soil will be subject to error. However, it seems likely that the amount of soil carbon is at least $4{\cdot}87 \times 10^{14}$ kilograms, almost forty times as much as the annual loss of carbon from the soil.

(*b*) *The Nitrogen Cycle*

Cyclical processes involving nitrogen differ from those involving carbon, as nitrogen gas, unlike carbon dioxide, is not directly metabolised by higher plants. It is fixed mostly by micro-organisms in the soil (p. 133) and then transferred to the plants. Nitrogen is returned to the soil in plant residues and is then released into the atmosphere following a variety of transformations, e.g. ammonification (p. 93), nitrification

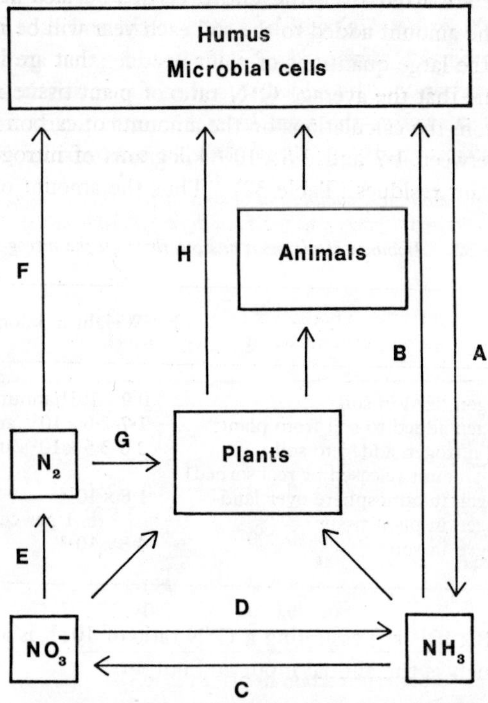

Fig. 27. *A generalised form of the nitrogen cycle.* A. *Ammonification and mineralisation.* B. *Immobilisation.* C. *Nitrification.* D. *Nitrate reduction and immobilisation.* E. *Denitrification.* F. *Nitrogen fixation (non-symbiotic).* G. *Nitrogen fixation (symbiotic).* H. *Uptake of nitrogen secreted from plants*

(*Reproduced, in modified form by permission of M. Alexander and John Wiley & Sons Inc.*)

(p. 157) and denitrification (p. 151): these reactions are summarised in Fig. 27.

The nitrogen cycle differs from the carbon cycle in other respects. Whereas carbon represents only 0·007 per cent of the atmosphere, nitrogen constitutes 80 per cent and so the reservoir of nitrogen is much greater. On the other hand, the percentage of nitrogen in plant tissue is much lower than that of carbon. Carbon:nitrogen ratios may be as high as 200:1 in living woody tissues, though they are much lower in legumes (30:1) and even lower in soil organic matter (10:1).

Rates of nitrogen fixation are low when compared with those of carbon fixation. It has been estimated that about $0{\cdot}9 \times 10^{11}$ kilograms of nitrogen are fixed over the entire earth's surface in one year[6]. However, the amount added to the soil each year will be much greater because of the large quantities of plant residues that are incorporated. If we assume that the average C:N ratio of plant tissue is 50:1, then it follows from the calculations on the amounts of carbon added to the soil, that between 1·7 and $2{\cdot}6 \times 10^{11}$ kilograms of nitrogen reach the soil from plant residues (Table 32). Thus the amount of nitrogen in

Table 32. *Amounts of nitrogen passing through the nitrogen cycle*

	Weight in kilograms
Nitrogen fixed in soil	$0{\cdot}9 \times 10^{11}$/annum
Nitrogen added to soil from plants	$1{\cdot}7\text{-}2{\cdot}6 \times 10^{11}$/annum
Total nitrogen added to soil (= amount released or reabsorbed)	$2{\cdot}6\text{-}3{\cdot}5 \times 10^{11}$/annum
Nitrogen in atmosphere over land	$1{\cdot}8 \times 10^{19}$
Nitrogen in plant tissue	? (*c.* 1 per cent)
Nitrogen in soil	$4{\cdot}8 \times 10^{13}$

the soil ($4{\cdot}8 \times 10^{13}$ kg) assuming a C:N ratio of 10:1, is about 150-200 times as much as the annual input and output.

(*c*) *The Phosphorus Cycle*

The circulation of phosphorus in nature, in the form of phosphate, is different from the circulation of both carbon and nitrogen. Phosphorus is almost entirely restricted to soils, rocks and living organisms and does not occur in the atmosphere. The total amount of phosphorus in the earth is large and in the crust alone there are about 2×10^{18} kg, i.e. about 0·12 per cent. This acts as a source of phosphorus for the soil where the content is lower, i.e. 0.04 per cent corresponding to about $1{\cdot}2 \times 10^{13}$ kg, mainly in the form of orthophosphates which may be organically or inorganically bound. Organically bound phosphate may constitute about 20-80 per cent of the total supply and is inert until attacked and mineralised by micro-organisms[261].

The inorganic phosphorus in soil is only sparingly soluble and is usually present in concentrations of 1-0·1 p.p.m. of the soil solution, where it exists as phosphoric acid (H_3PO_4), the corresponding ions $H_2PO_4^-$, HPO_4^{--} and PO_4^{---} and the complexes of these ions. As we saw in Chapter 7, in acid soils most of the non-complexed phosphate

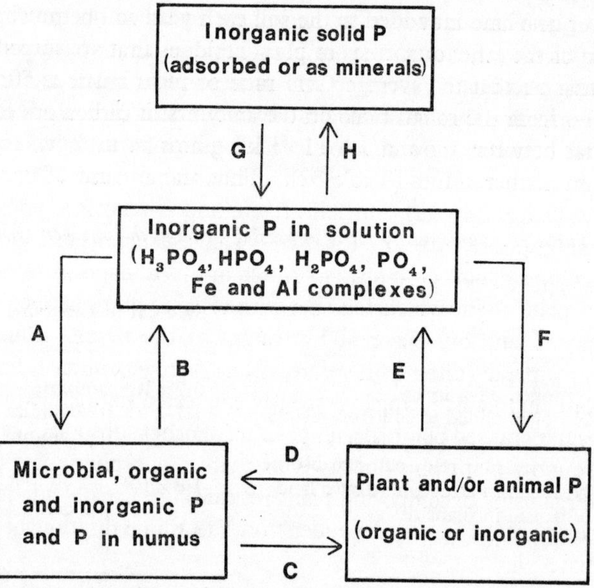

Fig. 28. *A generalised form of the phosphorus cycle.* A. *Microbial utilization of P.* B. *Lysis of micro-organisms and enzymic attack.* C. *P transfer, e.g. by mycorrhiza.* D. *P release from plant litter and animal remains by microbial action.* E. *P release from plant litter and animal remains by leaching.* F. *Direct uptake of P by plants or animals.* G. *Microbial solution of P and non-biological solution.* H. *Non-biological precipitation of P*

exists as $H_2PO_4^-$ ions, whilst in alkaline soils HPO_4^{--} ions predominate. Little is known about the phosphate complexes, although it is suspected that iron and aluminium complexes may be important in all soils and Larsen[261] has suggested that it is an oversimplification to assume that most of the soluble phosphorus is present in simple ionic form.

Microbial activity can influence the amount of phosphate in solution by immobilising it or by dissolving it (Fig. 28). Immobilisation of phosphate by micro-organisms (reaction A) is probably insignificant unless high microbial activity occurs, for in most soils the phosphate in the soil solution is replaced from phosphate in the solid phase, e.g. calcium phosphates. Similarly, solubilisation of phosphate by microbially produced acids (reaction G) is usually unimportant since the acids needed are rapidly degraded by other micro-organisms; possibly solubilisation through the accumulation of carbon dioxide and hydrogen sulphide is more important.

Solid phosphate may exist as phosphate adsorbed on the surface of soil particles, in the form of sparingly soluble minerals, e.g. apatite, and as organic phosphate. Organic phosphates are of much interest since they all originate through biological activity. It is difficult to establish from the literature how much inorganic phosphorus is taken up by plants from the soil and converted into organic forms, e.g. inositol hexaphosphate (phytin), nucleic acids and adenosine phosphates, although figures recorded by Russell[423] suggest that over 22 kg/hectare are taken up by some crop plants. Much of this remains in an inorganic form in plant tissue which together with organic phosphate constitutes between 0·5 and 5·0 per cent by weight of the tissue. Phosphorus uptake by plants differs, therefore, from nitrogen uptake, for in the latter case, inorganic nitrogen is mostly converted to organic forms and must be mineralised before decomposition of other substances can occur. Because most plant phosphorus is inorganic, phosphorus supplies for micro-organisms decomposing plant residues are usually adequate and so decomposition is not dependent on the mineralisation of organic phosphate[31].

The inorganic phosphorus in dead plant material is utilised by micro-organisms and converted into organic forms. The total phosphorus content of microbial cells varies from 0·5-1·0 per cent for fungi to 1·5-2·5 per cent for bacteria and actinomycetes[6]. This phosphorus only becomes available to plants and other micro-organisms when the micro-organisms containing it undergo autolysis and when enzymatic dephosphorylation occurs (p. 145). During decomposition of mature plant residues, little mineralisation of organic phosphorus occurs during the first three months, although it is possible that this process is masked by the simultaneous uptake of inorganic phosphorus by micro-organisms.

Thus, the overall result of plant decomposition seems to be that unchanged plant organic phosphorus and microbial organic phosphorus accumulate in the soil. These are subsequently decomposed by micro-organisms but at a rather slow rate[31]. It has been suggested that some bacteria will use organic phosphorus in preference to inorganic forms, although all the phosphorus mineralised by these reactions is used by the micro-organisms[32].

Control of Energy and Matter-flow in Soil

The way in which energy and matter pass through the soil-plant ecosystem is clearly important in relation to soil fertility. If it were possible to control parts of the pathways described, either by suppressing or encouraging them, then it might be possible to increase food production. So far, no real success has been achieved in this direction with respect to the saprophytic microbial populations of soil for the reasons outlined in our earlier considerations of nitrogen-fixation (p. 133), nitrification (p. 93) and solution of phosphate (p. 144). It remains more economical to continue addition of artificial fertilisers to soil, but whether this will always be so remains to be seen. Constant addition of ammonium fertilisers to soil has led to the build-up of nitrates in drainage water so that large bodies of water are now becoming increasingly polluted with this toxic substance. Any further increase in this already serious problem may result in a need to explore other methods of controlling plant growth, and it is in this connection that it is necessary to obtain more insight into the flow of energy in the soil. We have seen that the amounts of energy diverted into crop plants, animals and decomposition processes varies from one habitat to another and if we are to exploit our environment to the full, the reasons for these differences must be established.

REFERENCES

1 ABRAHAM, T. A. & HERR, L. J. 1964. Activity of actinomycetes from rhizosphere and non-rhizosphere soils of corn and soybean in four physiological tests. *Can. J. Microbiol.* **10**, 281–285.

2 ADAMS, M. E. & POSTGATE, J. R. 1960. On sporulation in sulphate-reducing bacteria. *J. gen. Microbiol.* **24**, 291–294.

3 ADAMS, M. H. 1959. *Bacteriophages.* Interscience, New York.

4 ADATI, M. 1939. Untersuchungen über die Rhizosphäre der Pflanzen. Zweiter Bericht. Über die Einflusse der verschiedenen Kulterpflanzen auf die Mikrobenzahl der verschiedenen Bodenarten in Formosa. *J. Soc. trop. Agric., Taiwan* **11**, 57–65.

5 AGATE, A. E. & BHAT, J. V. 1964. Microflora associated with the rhizosphere of *Calotropis gigantea. J. Ind. Inst. Sci.* **46**, 1–10.

6 ALEXANDER, M. 1961. *Introduction to soil microbiology.* Wiley, New York.

7 ALEXANDER, M. 1964. Biochemical ecology of soil micro-organisms. *A. Rev. Microbiol.* **18**, 217–252.

8 ALEXANDER, M. 1965. Biodegradation: problems of molecular recalcitrance and microbial fallibility. *Adv. appl. Microbiol.* **7**, 35–80.

9 ALEXANDER, M. 1968a. Degradation of pesticides by soil bacteria. In: *The ecology of soil bacteria.* Ed. by Gray, T. R. G. & Parkinson, D. 270–284. Liverpool University Press, Liverpool.

10 ALEXANDER, M. 1968b. Contribution to a discussion on the growth of bacteria in soil. In: *The ecology of soil bacteria.* Ed. by Gray, T. R. G. & Parkinson, D. 629. Liverpool University Press, Liverpool.

11 ALEXANDER, M. & CLARK, F. E. 1965. Nitrifying bacteria. In: *Methods of soil analysis* **2.** *Chemical and microbiological properties.* Ed. by Black, C. A. 1477–1483. American Society of Agronomy, Madison.

12 ALEXANDER, M. & WILSON, P. W. 1954. Large-scale production of the azotobacter for enzymes. *Appl. Microbiol.* **2**, 135–140.

13 ALLISON, F. E., HOOVER, S. R. & MORRIS, H. J. 1937. Physiological

studies with the nitrogen-fixing alga, *Nostoc muscorum*. *Bot. Gaz.* **98**, 433–463.

14 Allison, F. E. & Klein, C. J. 1962. Rates of immobilization and release of nitrogen following additions of carbonaceous materials and nitrogen to soils. *Soil Sci.* **93**, 383–386.

15 Aronson, J. M. 1965. The cell wall. In: *The fungi* **1**. Ed. by Ainsworth, G. C. & Sussman, A. S. 49–76. Academic Press, New York and London.

16 Bacon, J. S. D. 1968. The chemical environment of bacteria in soil. In: *The ecology of soil bacteria*. Ed. by Gray, T. R. G. & Parkinson, D. 23–43. Liverpool University Press, Liverpool.

17 Baker, J. M. 1963. Ambrosia beetles and their fungi, with particular reference to *Platypus cylindrus* Fab. In: *Symbiotic associations*. Ed. by Nutman, P. S. & Mosse, B. *Soc. gen. Microbiol. Symp.* **13**, 232–265. Cambridge University Press, Cambridge.

18 Barber, D. A. 1967. Influence of pH on uptake of phosphate by barley plants in sterile and non-sterile conditions. *Nature, Lond.* **215**, 779–780.

19 Barber, D. A. & Loughman, B. C. 1967. The effect of micro-organisms on the absorption of inorganic nutrients by intact plants. II. Uptake and utilization of phosphate by barley plants grown under sterile and non-sterile conditions. *J. exp. Bot.* **18**, 170–176.

20 de Barjac, H. & Chalvignac, M.A. 1954. Nouvel essai sur la determination du pouvoir amylolytique. *Annls Inst. Pasteur, Paris* **87**, 84–89.

21 Bartholomew, W. V. 1965. Mineralization and immobilization of nitrogen in the decomposition of plant and animal residues. In: *Soil nitrogen*. Ed. by Bartholomew, W. V. & Clark, F. E. 285–306. American Society of Agronomy, Madison.

22 Barton, R. 1960. Saprophytic activity of *Pythium mamillatum* in soils. I. Influence of substrate competition and soil environment. *Trans. Br. mycol. Soc.* **43**, 529–540.

23 Barton, R. 1961. Saprophytic activity of *Pythium mamillatum* in soils. II. Factors restricting *P. mamillatum* to pioneer colonization of substrates. *Trans. Br. mycol. Soc.* **44**, 105–118.

24 Basaraba, J. & Starkey, R. L. 1966. Effect of plant tannins on decomposition of organic substances. *Soil Sci.* **101**, 17–23.

25 Basu, S. N. & Ghose, S. N. 1960. The production of cellulase by fungi on mixed cellulosic substrates. *Can. J. Microbiol.* **6**, 265–282.

26 Becking, J. K. 1961. Studies on nitrogen-fixing bacteria of the

genus *Beijerinckia*. I. Geographical and ecological distribution in soils. *Pl. Soil* **14**, 49–81.

27 BECKING, J. H. 1968. Nitrogen fixation by non-leguminous plants. *Stikstof*, No. 12, 47–74.

28 BECKING, J. H., DE BOER, W. E. & HOUWINK, A. L. 1964. Electron microscopy of the endophyte of *Alnus glutinosa*. *Antonie van Leeuwenhoek* **30**, 343–376.

29 BERGERSEN, F. J. 1968. The symbiotic state in legume root nodules: studies with the soybean system. *Proc. 9th int. Congr. Soil Sci.* **2**, 49–63.

30 BERGERSEN, F. J. & BRIGGS, M. J. 1958. Studies on the bacterial component of soyabean root nodules; cytology and organisation of the host tissue. *J. gen. Microbiol.* **19**, 482–490.

31 BIRCH, H. F. 1961. Phosphorus transformations during plant decomposition. *Pl. Soil* **15**, 347–366.

32 BIRCH, H. F. 1964. The effect of 2:4 dinitro-phenol on phosphorus transformations during humus decomposition. *Pl. Soil* **21**, 391–394.

33 BLOOMFIELD, C. 1953. The mobilization of iron and aluminium by Scots pine needles. *J. Soil Sci.* **4**, 5–16.

34 BOCOCK, K. L. & GILBERT, O. J. W. 1957. The disappearance of leaf litter under different woodland conditions. *Pl. Soil* **9**, 179–185.

35 BOCOCK, K. L., GILBERT, O. J. W., CAPSTICK, C. K., TWINN, D. C., WAID, J. S. & WOODMAN, M. J. 1960. Changes in leaf litter when placed on the surface of soils with contrasting humus types. I. Losses in dry weight of oak and ash leaf litter. *J. Soil Sci.* **11**, 1–9.

36 BOND, G. 1963. The root nodules of non-leguminous angiosperms. In: *Symbiotic associations*. Ed. by Nutman, P. S. & Mosse, B. *Symp. Soc. gen. Microbiol.* **13**, 72–91. Cambridge University Press, Cambridge.

37 BOND, R. D. 1959. Occurrence of microbiological filaments in soils. *Nature, Lond.* **184**, 744–745.

38 BOOSALIS, M. G. & MANKAU, R. 1965. Parasitism and predation of soil micro-organisms. In: *Ecology of soil-borne plant pathogens*. Ed. by Baker, K. F. & Snyder, W. C. 374–389. University of California Press, Berkeley.

39 BOOTH, W. E. 1941. Algae as pioneers in plant succession and their importance in erosion control. *Ecology* **22**, 38–46.

40 BOUGHY, A. S., MUNRO, P. E., MEIKLEJOHN, J., STRANG, R. M. & SWIFT, J. M. 1964. Antibiotic reactions between African savanna species. *Nature, Lond.* **203**, 1302–1303.

41 Boulter, D., Jeremy, J. J. & Wilding, M. 1966. Amino acids liberated into the culture medium by pea seedling roots. *Pl. Soil* **24**, 121–127.

42 Bowen, H. J. M. & Cawse, P. A. 1964. Effects of ionizing radiation on soils and subsequent crop growth. *Soil Sci.* **97**, 252–259.

43 Braun, A. C. & Pringle, R. B. 1959. Pathogen factors in the physiology of disease—toxins and other metabolites. In: *Plant pathology, problems and progress 1908–1958*. Ed. by Holton, C. S. *et al.* 88–89. University of Wisconsin Press, Madison.

44 Bray, J. R. 1963. Root production and the estimation of net productivity. *Can. J. Bot.* **41**, 65–72.

45 Bremner, J. M. & Shaw, K. 1958. Denitrification in soil. II. Factors affecting denitrification. *J. agric. Sci.* **51**, 40–52.

46 Brian, P. W. 1957. The effects of some microbial metabolic products on plant growth. *Symp. Soc. exp. Biol.* **11**, 166–182.

47 Brian, P. W. 1960. Antagonistic and competitive mechanisms limiting survival and activity of fungi in soil. In: *The ecology of soil fungi*. Ed. by Parkinson, D. & Waid, J. S. 115–129. Liverpool University Press, Liverpool.

48 Broadbent, F. E. 1965. Effect of fertilizer nitrogen on the release of soil nitrogen. *Proc. Soil Sci. Soc. Am.* **29**, 692–696.

49 Broadbent, F. E., Tyler, K. B. & Hill, G. N. 1958. Nitrification of fertilizers. *Calif. Agric.* **12**, (10) 9, 14.

50 Brock, T. D. 1966. *Principles of microbial ecology*. Prentice-Hall, New Jersey.

51 Brock, T. D. 1967a. Bacterial growth rate in the sea: direct analysis by thymidine autoradiography. *Science* **155**, 81–83.

52 Brock, T. D. 1967b. The ecosystem and the steady state. *Bioscience* **17**, 166–169.

53 Brooks, M. A. 1963. Symbiosis and aposymbiosis in arthropods. In: *Symbiotic associations*. Ed. by Nutman, P. S. & Mosse, B. *Symp. Soc. gen. Microbiol.* **13**, 200–231. Cambridge University Press, Cambridge.

54 Brown, M. E., Burlingham, S. K. & Jackson, R. M. 1962. Studies on *Azotobacter* species in soil. I. Comparison of media and techniques for counting *Azotobacter* in soil. *Pl. Soil* **17**, 309–319.

55 Brown, M. E., Jackson, R. M. & Burlingham, S. K. 1968. Growth and effects of bacteria introduced into soil. In: *The ecology of soil bacteria*. Ed. by Gray, T. R. G. & Parkinson, D. 531–551. Liverpool University Press, Liverpool.

56 BUDDENHAGEN, I. W. 1965. The relation of plant pathogenic bacteria to the soil. In: *Ecology of soil-borne plant pathogens.* Ed. by Baker, K. F. & Snyder, W. C. 269–284. University of California Press, Berkeley.

57 BURGES, A. 1950. The downward movement of fungal spores in sandy soil. *Trans. Br. mycol. Soc.* **33**, 142–147.

58 BURGES, A. 1958. *Micro-organisms in the soil.* Hutchinson, London.

59 BURGES, A. 1960. Dynamic equilibria in the soil. In: *The ecology of soil fungi.* Ed. by Parkinson, D. & Waid, J. S. 185–191. Liverpool University Press, Liverpool.

60 BURGES, A. 1965. Biological processes in the decomposition of organic matter. In: *Experimental pedology.* Ed. by Hallsworth, E. G. & Crawford, D. V. 189–198. Butterworths, London.

61 BURGES, A., HURST, H. M. & WALKDEN, B. 1964. The phenolic constituents of humic acid and their relation to the lignin of the plant cover. *Geochim. cosmochim. Acta* **28**, 1547–1554.

62 BURGES, A. & LATTER, P. 1960. Decomposition of humic acid by fungi. *Nature, Lond.* **186**, 404–405.

63 BURGES, A. & NICHOLAS, D. P. 1961. Use of soil sections in studying amounts of fungal hyphae in soil. *Soil Sci.* **92**, 25–29.

64 BURNETT, J. H. 1968. *Fundamentals of mycology.* Arnold, London.

65 BURNS, G. R. 1967. Oxidation of sulphur in soils. *Tech. Bull. Sulphur Inst.* **13**, 1–41.

66 BURSTROM, H. G. 1965. The physiology of plant roots. In: *Ecology of soil-borne plant pathogens.* Ed. by Baker, K. F. & Snyder, W. C. 154–167. University of California Press, Berkeley.

67 CALDWELL, R. 1958. Fate of spores of *Trichoderma viride* Pers. ex. Fr. introduced into soil. *Nature, Lond.* **181**, 1144–1145.

68 CARLILE, M. J. 1956. A study of the factors influencing non-genetic variation in a strain of *Fusarium oxysporum. J. gen. Microbiol.* **14**, 643–654.

69 CASIDA, L. E. JR. 1962. On the isolation and growth of individual microbial cells from soil. *Can. J. Microbiol.* **8**, 115–119.

70 CASIDA, L. E. JR. 1965. Abundant micro-organism in soil. *Appl. Microbiol.* **13**, 327–334.

71 CASIDA, L. E. JR., KLEIN, D. A. & SANTORO, T. 1964. Soil dehydrogenase activity. *Soil Sci.* **98**, 371–376.

72 CASTELL, C. H. & GARRARD, E. M. 1941. The action of micro-organisms on fat. II. Oxidation and hydrolysis of triolein by pure cultures of bacteria. *Can. J. Res.* **C19**, 106–110.

73 CHAHAL, K. S. & WAGNER, G. H. 1965. Decomposition of organic matter in Sanborn field soils amended with C14 glucose. *Soil Sci.* **100**, 96–103.

74 CHALVIGNAC, M. A. 1953. Mesure des pouvoirs amylolytique et protéolytiques des terres en aérobiose. *Annls Inst. Pasteur, Paris* **84**, 816–819.

75 CHAN, E. C. S. & KATZNELSON, H. 1961. Growth of *Arthrobacter globiformis* and *Pseudomonas* sp. in relation to the rhizosphere effect. *Can. J. Microbiol.* **7**, 759–767.

76 CHAN, E. C. S., KATZNELSON, H. & ROUATT, J. W. 1963. The influence of soil and root extracts on the associative growth of selected soil bacteria. *Can. J. Microbiol.* **9**, 187–197.

77 CHAPMAN, H. D. 1965. Chemical factors of the soil as they affect micro-organisms. In: *Ecology of soil-borne plant pathogens.* Ed. by Baker, K. F. & Snyder, W. C. 120–139. University of California Press, Berkeley.

78 CHASE, F. E., CORKE, C. T. & ROBINSON, J. B. 1968. Nitrifying bacteria in soil. In: *The ecology of soil bacteria.* Ed. by Gray, T. R. G. & Parkinson, D. 591–611. Liverpool University Press, Liverpool.

79 CHESHIRE, M. V. & MUNDIE, C. M. 1966. The hydrolytic extraction of carbohydrates from soil by sulphuric acid. *J. Soil. Sci.* **17**, 372–381.

80 CHESTERS, C. G. C. 1940. A method of isolating soil fungi. *Trans. Br. mycol. Soc.* **24**, 352–355.

81 CHESTERS, C. G. C. 1948. A contribution to the study of fungi in the soil. *Trans. Br. mycol. Soc.* **30**, 100–117.

82 CHET, I., HENIS, Y. & MITCHELL, R. 1967. Chemical composition of hyphal and sclerotial walls of *Sclerotium rolfsii* Sacc. *Can. J. Microbiol.* **13**, 137–141.

83 CHINN, S. H. F. & LEDINGHAM, R. J. 1961. Mechanisms contributing to the eradication of spores of *Helminthosporium sativum* from amended soil. *Can. J. Microbiol.* **39**, 739–748.

84 CHOLODNEY, N. G. 1930. Über eine neue Methode zur Untersuchung der Bodenmikroflora. *Arch. Mikrobiol.* **1**, 620–652.

85 CLARK, C. & SCHMIDT, E. L. 1966. Effect of mixed culture on *Nitrosomonas europaea* simulated by uptake and utilization of pyruvate. *J. Bact.* **91**, 367–373.

86 CLARK, F. E. 1967. Bacteria in soil. In: *Soil biology.* Ed. by Burges, A. & Raw, F. 15–49. Academic Press, London and New York.

87 CLARK, F. E. 1968. The growth of bacteria in soil. In: *The ecology of soil bacteria.* Ed. by Gray, T. R. G. & Parkinson, D. 441–457. Liverpool University Press, Liverpool.

88 CLARKE, P. H. & LILLY, M.D. 1969. The regulation of enzyme synthesis during growth. In: *Microbial growth.* Ed. by Meadow, P. & Pirt, S. J. *Symp. Soc. gen. Microbiol.* **19**, 113–159. Cambridge University Press, Cambridge.

89 CLOWES, F. A. L & JUNIPER, B. E. 1968. *Plant cells.* Blackwells, Oxford.

90 COCHRANE, V. W. 1958. *Physiology of fungi.* Wiley, New York & London.

91 CONN, H. J. 1918. The microscopic study of bacteria and fungi in soil. *Tech. Bull. N.Y. agric. exp. Sta.* **64.**

92 COOK, F. D. & LOCHHEAD, A. G. 1959. Growth factor relationships of soil micro-organisms as affected by proximity to the plant root. *Can. J. Microbiol.* **5**, 323–334.

93 COOKE, R. C. 1962. Behaviour of nematode-trapping fungi during decomposition of organic matter in soil. *Trans Br. mycol. Soc.* **45**, 314–320.

94 COOPER, R. 1959. Bacterial fertilizers in the Soviet Union. *Soils Fertil., Harpenden* **22**, 327–333.

95 CORKE, C. T. & CHASE, F. E. 1964. Comparative studies of actinomycete populations in acid podzolic and neutral mull forest soils. *Proc. Soil Sci. Soc. Am.* **28**, 68–70.

96 COUDERCHET, J. 1967. Action du basidiomycète *Rhodopaxillus saevus* (d'un rond de Sorcière) sur la microflore tellurique. *Revue gén. Bot.* **74**, 107–134.

97 COULSON, C. B., DAVIES, R. I. & LEWIS, D. A. 1960. Polyphenols in plant, humus and soil. I. Polyphenols of leaves, litter and superficial humus from mull and mor sites. *J. Soil Sci.* **11**, 20–29.

98 CROSS, T., WALKER, P. D. & GOULD, G. W. 1968. Thermophilic actinomycetes producing resistant endospores. *Nature, Lond.* **220**, 352–354.

99 CROSSE, J. E. 1968. Plant pathogenic bacteria in soil. In: *The ecology of soil bacteria.* Ed. by Gray, T. R. G. & Parkinson D. 552–572. Liverpool University Press, Liverpool.

100 CURRIE, J. A. 1962. The importance of aeration in producing the right conditions for plant growth. *J. Sci. Fd Agric.* **13**, 380–385.

101 DAFT, M. J. & NICHOLSON, T. H. 1966. Effect of *Endogone* mycorrhiza on plant growth. *New Phytol.* **65**, 343–350.

102 DARKEN, M. A. 1962. Absorption and transport of fluorescent brighteners by micro-organisms. *Appl. Microbiol.* **10**, 387–393.

103 DART, P. J. & MERCER, F. V. 1963. Development of the bacteroid in the root nodule of Barrel medic (*Medicago tribuloides* Desr.)

and subterranean clover (*Trifolium subterraneum* L.). *Arch. Mikrobiol.* **46**, 382–401.

104 DART, P. J. & MERCER, F. V. 1963. Membrane envelopes of legume nodule bacteroids. *J. Bact.* **85**, 951–952.

105 DART, P. J. & MERCER, F. V. 1964. The legume rhizosphere. *Arch. Mikrobiol.* **47**, 344–378.

106 DAVEY, C. B. & WILDE, S. A. 1955. Determination of the numbers of soil micro-organisms by the use of molecular membrane filters. *Ecology* **36**, 760–761.

107 DAVIES, R. I., COULSON, C. B. & LEWIS, D. A. 1964. Polyphenols in plant, humus and soil. IV. Factors leading to increase in biosynthesis of polyphenol in leaves and their relationship to mull and mor formation. *J. Soil Sci.* **15**, 310–318.

108 DE, P. K. & MANDAL, L. N. 1956. Fixation of nitrogen by algae in rice soils. *Soil Sci.* **81**, 453–458.

109 DELONG, W. A. & SCHNITZER, M. 1955. Investigations on the mobilization and transport of iron in forested soils. I. The capacities of leaf extracts and leachates to react with iron. *Proc. Soil Sci. Soc. Am.* **19**, 360–363.

110 DICK, M. W. 1963. The occurrence of Saprolegniaceae in certain soils of southeast England. III. Distribution in relation to pH and water content. *J. Ecol.* **51**, 75–82.

111 DICKEY, R. S. 1961. Relation of some edaphic factors to *Agrobacterium tumefaciens*. *Phytopathology* **51**, 607–614.

112 DICKINSON, C. H. & PUGH, G. J. F. 1965. The mycoflora associated with *Halimione portulacoides*. II. Root surface fungi of mature and excised plants. *Trans. Br. mycol. Soc.* **48**, 595–602.

113 DIX, N. J. 1964. Colonization and decay of bean roots. *Trans. Br. mycol. Soc.* **47**, 285–292.

114 DOBBS, C. G. & GASH, M. J. 1965. Microbial and residual mycostasis in soils. *Nature, Lond.* **207**, 1354–1356.

115 DOBBS, C. G. & HINSON, W. H. 1953. A widespread fungistasis in soils. *Nature, Lond.* **175**, 500–501.

116 DOMMERGUES, Y. 1959. L'activité de la microflore tellurique aux faibles humidités. *C. r. Acad. Sci., Paris* **248**, 487–490.

117 DOMMERGUES, Y. 1960. Mineralization de l'azote aux faibles humidités. *Trans. 7th int. Congr. Soil Sci.* **2**, 672–678.

118 DUDDINGTON, C. L. 1957a. *The friendly fungi. A new approach to the eelworm problem.* Faber and Faber, London.

119 DUDDINGTON, C. L. 1957b. The predacious fungi and their place in microbial ecology. In: *Microbial ecology.* Ed. by Spicer,

C. C. & Williams, R. E. O. *Symp. Soc. gen. Microbiol.* **7**, 218–237. Cambridge University Press, Cambridge.

120 DUDDINGTON, C. L. 1963. Predacious fungi and soil nematodes. In: *Soil organisms.* Ed. by Doeksen, J. & van der Drift, J. 298–304. North Holland Publishing Co., Amsterdam.

121 DUFF, R. B., WEBLEY, D. M. & SCOTT, R. O. 1963. Solubilization of minerals and related materials by 2-ketogluconic acid-producing bacteria. *Soil Sci.* **95**, 105–114.

122 EDDY, B. P. 1960. The use and meaning of the term 'psychrophilic'. *J. appl. Bact.* **23**, 189–190.

123 EDWARDS, C. A. & HEATH, G. W. 1963. The role of soil animals in breakdown of leaf litter. In: *Soil organisms.* Ed. by Doeksen, J. & van der Drift, J. 76–84. North Holland Publishing Co., Amsterdam.

124 EGGINS, H. O. W. & PUGH, G. J. F. 1961. Isolation of cellulose-decomposing fungi from the soil. *Nature, Lond.* **193**, 94–95.

125 ELWAN, S. H. & MAHMOUD, S. A. Z. 1960. Note on the bacterial flora of the Egyptian desert in summer. *Arch. Mikrobiol.* **36**, 360–364.

126 ENO, C. F. & POPENOE, H. 1964. Gamma radiation compared with steam and methyl bromide as a soil sterilizing agent. *Proc. Soil Sci. Soc. Am.* **28**, 533–535.

127 EREN, J. & PRAMER, D. 1965. The most probable number of nematode-trapping fungi in soil. *Soil Sci.* **99**, 285.

128 EREN, J. & PRAMER, D. 1966. Application of immunofluorescent staining to studies of the ecology of soil micro-organisms. *Soil Sci.* **101**, 39–45.

129 ESTERMANN, E. F., PETERSON, G. H. & MCLAREN, A. D. 1959. Digestion of clay-protein, lignin-protein, and silica-protein complexes by enzymes and bacteria. *Proc. Soil Sci. Soc. Am.* **23**, 31–36.

130 FÅHRAEUS, G. & LJUNGGREN, H. 1959. The possible significance of pectic enzymes in root hair infection by nodule bacteria. *Physiol. Plant.* **12**, 145–154.

131 FÅHRAEUS, G. & LJUNGGREN, H. 1968. Pre-infection phases of the legume symbiosis. In: *The ecology of soil bacteria.* Ed. by Gray, T. R. G. & Parkinson, D. 396–421. Liverpool University Press, Liverpool.

132 FAY, P., STEWART, W. D. P., WALSBY, A. E. & FOGG, G. E. 1968. Is the heterocyst the site of nitrogen fixation in blue-green algae? *Nature, Lond.* **220**, 810–812.

133 FLENTJE, N. T. 1965. Pathogenesis by soil fungi. In: *Ecology of soil-borne plant pathogens.* Ed. by Baker, K. F. & Snyder, W. C. 255–266. University of California Press, Berkeley.

134 Fogg, G. E. 1951. Growth and heterocyst production in *Anabaena cylindrica* Lemn. III. The cytology of heterocysts. *Ann. Bot., Lond.* NS **15**, 23–35.

135 Fried, M. & Broeshart, H. 1967. *The soil-plant system in relation to inorganic nutrition.* Academic Press, New York and London.

136 Fritsch, F. E. 1945. *The structure and reproduction of the algae,* **2**. Cambridge University Press, Cambridge.

137 Furman, T. E. 1959. The structure of the root nodules of *Ceanothus sanguineus* and *C. velutinus,* with special reference to the endophyte. *Am. J. Bot.* **46**, 698–703.

138 Gadgil, P. D. 1963. Soil sections of grassland. In: *Soil organisms.* Ed. by Doeksen, J. & van der Drift, J. 327–332. North Holland Publishing Co., Amsterdam.

139 Gainey, P. L. 1930. A study of factors influencing inoculation experiments with *Azotobacter. Tech. Bull. Kans. agric. Exp. Sta.* **26**, 3.

140 Garrett, S. D. 1950. Ecology of the root inhabiting fungi. *Biol. Rev.* **25**, 220–254.

141 Garrett, S. D. 1951. Ecological groups of soil fungi; a survey of substrate relationships. *New Phytol.* **50**, 149–166.

142 Garrett, S. D. 1956. *Biology of root infecting fungi.* Cambridge University Press, Cambridge.

143 Garrett, S. D. 1963. *Soil fungi and soil fertility.* Pergamon Press, Oxford.

144 Gaümann, E. & Kern, H. 1959. Über die Isolierung und den chemischen Nachweis des Orchinols. *Phytopath. Z.* **36**, 347–356.

145 Gause, G. F. 1934. *The struggle for existence.* Williams and Wilkins, Baltimore.

146 Georgieva, J. & Sheikova, G. 1963. Vitamin B_{12} formation from various actinomycetes isolated from Bulgarian soils. *Folia microbiol., Praha* **8**, 322–324.

147 Gerretsen, F. C. 1948. The influence of micro-organisms on the phosphate intake by the plant. *Pl. Soil* **1**, 51–81.

148 Gledhill, W. E. & Casida, L. E. Jr. 1969. Predominant catalase-negative bacteria. I. Streptococcal population indigenous to soil. *Appl. Microbiol.* **17**, 208–213.

149 Gorham, E. 1959. Chlorophyll derivatives in woodland soils. *Soil Sci.* **87**, 258–261.

150 Gorham, E. & Sanger, J. 1967. Plant pigments in woodland soils. *Ecology* **48**, 306–308.

151 Goring, C. I. A. 1962. Control of nitrification by 2-chloro-6 (trichloromethyl) pyridine. *Soil Sci.* **93**, 211–218.

152 GOTTLIEB, D. & SIMINOFF, P. 1952. The production and role of antibiotics in the soil. II. Chloromycetin. *Phytopathology*, **42**, 91–97.

153 GRAY, L. E. & GERDEMANN, J. W. 1967. Influence of vesicular-arbuscular mycorrhizas on the uptake of phosphorus 32 by *Liriodendron tulipifera* and *Liquidambar styraciflua*. *Nature, Lond.* **213**, 106–107.

154 GRAY, T. R. G. 1967. Stereoscan electron microscopy of soil micro-organisms. *Science, NY.* **155**, 1668–1670.

155 GRAY, T. R. G. & BAXBY, P. 1968. Chitin decomposition in soil. II. The ecology of chitinoclastic micro-organisms in forest soil. *Trans. Br. mycol. Soc.* **51**, 293–309.

156 GRAY, T. R. G., BAXBY, P., HILL, I. R. & GOODFELLOW, M. 1968. Direct observation of bacteria in soil. In: *The ecology of soil bacteria*. Ed. by Gray, T. R. G. & Parkinson, D. 171–197. Liverpool University Press, Liverpool.

157 GRAY, T. R. G. & BELL, T. F. 1963. The decomposition of chitin in an acid soil. In: *Soil organisms*. Ed. by Doeksen, J. & van der Drift, J. 222–230. North Holland Publishing Co., Amsterdam.

158 GRAY, T. R. G. & LOWE, W. E. 1967. Techniques for studying cutin decomposition in soil. *Bact. Proc.*, 3.

159 GREAVES, M. P. & WEBLEY, D. M. 1965. A study of the breakdown of organic phosphate by micro-organisms from the root region of certain pasture grasses. *J. appl. Bact.* **28**, 454–465.

160 GREENLAND, D. J., LINDSTROM, G. R. & QUIRK, J. P. 1962. Organic materials which stabilize natural soil aggregates. *Proc. Soil Sci. Soc. Am.* **26**, 366–371.

161 GREENWOOD, D. J. 1961. The effect of oxygen concentration on the decomposition of organic materials in soil. *Pl. Soil* **24**, 360–376.

162 GREENWOOD, D. J. 1962. Nitrification and nitrate dissimilation in soil. II. Effect of oxygen concentration. *Pl. Soil* **17**, 378–391.

163 GREENWOOD, D. J. 1968. Measurement of microbial metabolism in soil. In: *The ecology of soil bacteria*. Ed. by Gray, T. R. G. & Parkinson, D. 138–157. Liverpool University Press, Liverpool.

164 GREENWOOD, D. J. & BERRY, G. 1962. Aerobic respiration in soil crumbs. *Nature, Lond.* **195**, 161–163.

165 GREENWOOD, D. J. & LEES, H. 1960. Studies on the decomposition of amino acids in soils. II. The anaerobic metabolism. *Pl. Soil* **12**, 69–80.

166 GRIFFIN, D. M. 1960. Fungal colonization of sterile hair in contact with soil. *Trans. Br. mycol. Soc.* **43**, 583-596.
167 GRIFFIN, D. M. 1963a. Soil physical factors and the ecology of fungi. I. Behaviour of *Curvularia ramosa* at small water suctions. *Trans. Br. mycol. Soc.* **46**, 273-280.
168 GRIFFIN, D. M. 1963b. Soil physical factors and the ecology of fungi. II. Behaviour of *Pythium ultimum* at small soil water suctions. *Trans. Br. mycol. Soc.* **46**, 368-372.
169 GRIFFIN, D. M. 1963c. Soil physical factors and the ecology of fungi. III. Activity of fungi in relatively dry soil. *Trans. Br. mycol. Soc.* **46**, 373–377.
170 GRIFFIN, D. M. 1963d. Soil moisture and the ecology of soil fungi. *Biol. Rev.* **38**, 141–166.
171 GRIFFIN, D. M. 1966. Soil physical factors and the ecology of fungi. IV. Influence of the soil atmosphere. *Trans. Br. mycol. Soc.* **49**, 115-119.
172 GRIFFIN, D. M. 1968. Observations on fungi growing in a translucent particulate matrix. *Trans. Br. mycol. Soc.* **51**, 319–322.
173 GRIFFITHS, D. A. 1966. Vertical distribution of mycostasis in Malayan soils. *Can. J. Microbiol.* **12**, 149-164.
174 GRIFFITHS, E. 1965. Micro-organisms and soil structure. *Biol. Rev.* **40**, 129–142.
175 GRIFFITHS, E. & BIRCH, H. F. 1961. Microbiological changes in freshly moistened soil. *Nature, Lond.* **189**, 424.
176 GRIFFITHS, E. & JONES, D. 1963. Colonization of cellulose by soil micro-organisms. *Trans. Br. mycol. Soc.* **46**, 285–294.
177 GUITTONEAU, G. 1927. Sur l'oxydation microbienne du soufre en cours de l' ammonisation. *C. r. Acad. Sci., Paris* **184**, 45–46.
178 GUNNER, H. B., ZUCKERMAN, B. M., WALKER, R. W., MILLER, C. W., DEUBERT, K. H. & LONGLEY, R. E. 1966. The distribution and persistence of diazinon applied to plant and soil and its influence on rhizosphere and soil microflora. *Pl. Soil* **25**, 249–264.
179 GUPTA, U. C. 1967. Carbohydrates. In: *Soil biochemistry*. Ed. by McLaren, A. D. & Peterson, G. H. 91–118. Arnold, London.
180 GUPTA, U. C., SOWDEN, F. J. & STOBBE, P. C. 1963. The characterization of carbohydrate constituents from different soil profiles. *Proc. Soil Sci. Soc. Am.* **27**, 380–382.
181 HAARLØV, N. & WEIS-FOGH, T. 1953. A microscopical technique for studying the undisturbed texture of soils. *Oikos* **4**, 44–57.
182 HADFIELD, W. 1960. Rhizosphere effect on soil algae. *Nature, Lond.* **185**, 178-179.

183 HANDLEY, W. R. C. 1954. Mull and mor formation in relation to forest soils. *Forestry Commission Bulletin* **23**.

184 HANDLEY, W. R. C. 1961. Further evidence for the importance of residual leaf protein complexes in litter decomposition and the supply of nitrogen for plant growth. *Pl. Soil* **15**, 37–73.

185 HARLEY, J. L. 1960. The physiology of soil fungi. In: *The ecology of soil fungi.* Ed. by Parkinson, D. & Waid, J. S. 265–276. Liverpool University Press, Liverpool.

186 HARLEY, J. L. 1965. Mycorrhiza. In: *Ecology of soil-borne plant pathogens.* Ed. by Baker, K. F. & Snyder, W. C. 218–229. University of California Press, Berkeley.

187 HARLEY, J. L. & WAID, J. S. 1955. A method of studying active mycelia on living roots and other surfaces in the soil. *Trans. Br. mycol. Soc.* **38**, 104–118.

188 HARLEY, J. L. & WILSON, J. M. 1959. The absorption of potassium by beech mycorrhiza. *New Phytol.* **58**, 281-298.

189 HARRIS, J. R. 1966. The transformation of nitrogen and production of phytotoxins during the decomposition of buried wheat straw. In: *9th Int. Congr. Microbiol., Abstracts.* 267. Ed. by Timakov, V. D. *et al.*, Moscow.

190 HARRIS, R. F., ALLEN, O. N., CHESTERS, G. & ATTOE, O. J. 1963. Evaluation of microbial activity in soil aggregate stabilization and degradation by the use of artificial aggregates. *Proc. Soil Sci. Soc. Am.* **27**, 542–545.

191 HARRISON, A. P. & LAWRENCE, F. R. 1963. Phenotypic, genotypic and chemical changes in starving populations of *Aerobacter aerogenes*. *J. Bact.* **85**, 742–750.

192 HART, M. G. R. 1959. Sulphur oxidation in tidal mangrove soils of Sierra Leone. *Pl. Soil* **11**, 215–236.

193 HART, M. G. R. 1963. Observations on the source of acid in empoldered mangrove soils. II. Oxidation of soil polysulphides. *Pl. Soil* **19**, 106–114.

194 HAWKER, L. E. 1957. *The physiology of reproduction in fungi.* Cambridge University Press, Cambridge.

195 HAWKER, L. E. 1962. Studies on vesicular-arbuscular endophytes. V. A review of the evidence relating to the identity of the causal fungus. *Trans. Br. mycol. Soc.* **45**, 190–199.

196 HAWKER, L. E., LINTON, A. H., FOLKES, B. F. & CARLILE, M. J. 1960. *An introduction to the biology of micro-organisms.* Arnold, London.

197 HAYES, A. J. 1965. Studies on the decomposition of coniferous leaf litter. I. Physical and chemical changes. *J. Soil Sci.* **16**, 121–140.

198 HEAL, O. W. 1970. Methods of study of soil protozoa. In: *Methods for the study of production and energy flow in soil communities.* Ed. by Phillipson, J. Unesco, Paris.

199 HEARD, A. J. 1965. The effect of the nitrogen content of residues from leys on amounts of available soil nitrogen and on yields of wheat. *J. agric. Sci.* **64,** 329–334.

200 HEINEN, W. 1963a. Über den enzymatischen cutin-abbau. V. Mitteilung: die lyse von peroxyd-brücken in cutin durch eine peroxydase aus *Penicillium spinulosum.* Thom. *Acta. bot. neerl.* **12,** 51–57.

201 HEINEN, W. 1963b. Enzymatische Aspekte zur Biosynthese des Blatt-Cutins bei *Gasteria verricuosa*-Blattern nach Verletzung. *Z. Naturf.* **18b,** 67–79.

202 HEINEN, W. & DE VRIES, H. 1966. Stages during the breakdown of plant cutin by soil micro-organisms. *Arch. Mikrobiol.* **54,** 331–338.

203 HENDERSON, M. E. K. 1961a. Isolation, identification and growth of some soil hyphomycetes and yeast-like fungi which utilize aromatic compounds related to lignin. *J. gen. Microbiol.* **26,** 149–154.

204 HENDERSON, M. E. K. 1961b. The metabolism of aromatic compounds related to lignin by some hyphomycetes and yeast-like fungi of soil. *J. gen. Microbiol.* **26,** 155–165.

205 HERING, T. F. 1965. Succession of fungi in the litter of a Lake District oakwood. *Trans. Br. mycol. Soc.* **48,** 391–408.

206 HESSE, P. R. 1959. Sulphur and nitrogen changes in forest soils of East Africa. *Pl. Soil* **9,** 86–96.

207 HESSELTINE, C. W., BRADLE, B. J. & BENJAMIN, C. R. 1960. Further investigations on the preservation of molds. *Mycologia* **52,** 762–774.

208 HILL, I. R. 1967. *Application of the fluorescent antibody technique to an ecological study of bacteria in soil.* Ph.D. thesis, University of Liverpool.

209 HILL, I. R. & GRAY, T. R. G. 1967. Application of the fluorescent-antibody technique to an ecological study of bacteria in soil. *J. Bact.* **93,** 1888–1896.

210 HIRST, J. M. & STEDMAN, O. J. 1963. Dry liberation of fungus spores by raindrops. *J. gen. Microbiol.* **33,** 335–344.

211 HITCHCOCK, A. E. & ZIMMERMAN, P. W. 1953. Absorption and movement of synthetic growth substances from soil as indicated by the responses of aerial parts. *Contrib. Boyce Thompson Inst.* **7,** 447–476.

212 HOLDING, A. J. 1960. The properties and classification of pre-

dominant Gram-negative bacteria occurring in soil. *J. appl. Bact.* **23**, 515–525.

213 Holding, A. J. & Jeffrey, D. C. 1968. Effects of metallic ions on soil bacteria. In: *The ecology of soil bacteria.* Ed. by Gray, T. R. G. & Parkinson, D. 516–530. Liverpool University Press, Liverpool.

214 Holding, A. J. & King, J. 1963. The effectiveness of indigenous populations of *Rhizobium trifolii* in relation to soil factors. *Pl. Soil* **18**, 191–198.

215 Holland, A. A. & Parker, C. A. 1966. Studies on microbial antagonism in the establishment of clover pasture. II. The effect of saprophytic soil fungi upon *Rhizobium trifolii* and the growth of subterranean clover. *Pl. Soil* **25**, 329–340.

216 Hopkins, F. G. & Cole, S. W. 1903. A contribution to the chemistry of proteids. II. The constitution of tryptophane, and the action of bacteria upon it. *J. Physiol. Lond.* **29**, 451–466.

217 Hoyt, P. B. 1966a. Chlorophyll-type compounds in soil. I. Their origin. *Pl. Soil* **25**, 167–180.

218 Hoyt, P. B. 1966b. Chlorophyll-type compounds in soil. II. Their decomposition. *Pl. Soil* **25**, 313–328.

219 Huber, D. M. & Anderson, A. L. 1966. Necrosis of hyphae of *Fusarium solani* f. *phaseoli* and *Rhizoctonia solani* induced by a soil-borne bacterium. *Phytopathology* **56**, 1416–1417.

220 Hungate, R. E. 1962. Ecology of bacteria. In: *The bacteria* **4**, *The physiology of growth.* Ed. by Gunsalus, I. C. & Stanier, R. Y. 95–119. Academic Press, New York & London.

221 Hurst, H. M. & Burges, N. A. 1967. Lignin and humic acids. In: *Soil biochemistry.* Ed. by McLaren, A. D. & Peterson, G. H. 260–286. Arnold, London.

222 Hurst, H. M., Burges, N. A. & Latter, P. M. 1962. Some aspects of the biochemistry of humic acid decomposition by fungi. *Phytochemistry* **1**, 227–231.

223 Jacks, G. V. 1965. The role of organisms in the early stages of soil formation. In: *Experimental pedology.* Ed. by Hallsworth, E. G. & Crawford, D. V. 219–226. Butterworth, London.

224 Jackson, R. M. 1958. An investigation of fungistasis in Nigerian soils. *J. gen. Microbiol.* **18**, 248–258.

225 Jackson, R. M. 1960. Soil fungistasis and the rhizosphere. In: *The ecology of soil fungi.* Ed. by Parkinson, D. & Waid. J. S. 168–176. Liverpool University Press, Liverpool.

226 Jacob, F. & Monod, J. 1961. Genetic regulatory mechanisms in the synthesis of proteins. *J. molec. Biol.* **3**, 318–356.

227 JAMES, N. 1958. Soil extract in soil microbiology. *Can. J. Microbiol.* **4**, 363-370.

228 JAMES, N. 1959. Plate counts of bacteria and fungi in a saline soil. *Can. J. Microbiol.* **5**, 431–439.

229 JEFFREYS, E. G., BRIAN, P. W., HEMMING, H. G. & LOWE, D. 1953. Antibiotic production by the microfungi of acid heath soils. *J. gen. Microbiol.* **9**, 314–341.

230 JENKINSON, D. S. 1965. Studies on the decomposition of plant material in soil. I. Loss of carbon from ^{14}carbon labelled ryegrass incubated with soil in the field. *J. Soil Sci.* **16**, 104–115.

231 JENKINSON, D. S. 1966. Studies on the decomposition of plant material in soil. II. Partial sterilization and the soil biomass. *J. Soil Sci.* **17**, 280–302.

232 JENNY, H. & GROSSENBACHER, K. 1962. Root-soil boundary zones. *Calif. Agric.* **16**, 7.

233 JENNY, H. & GROSSENBACHER, K. 1963. Root-soil boundary zones as seen by the electron microscope. *Proc. Soil Sci. Soc. Am.* **27**, 273–277.

234 JENSEN, H. L. 1958. The classification of the rhizobia. In: *Nutrition of the legumes.* Ed. by Hallsworth, E. G. 75–86. Butterworth, London.

235 JENSEN, V. 1962. Studies on the microflora of Danish beech forest soils. I. The dilution plate count technique for the enumeration of bacteria and fungi in soil. *Zentbl. Bakt. Parasit-Kde., Abt II*, **116**, 13–32.

236 JINKS, J. L. 1952. Heterocaryosis in wild *Penicillium. Heredity*, **6**, 77–87.

237 JINKS, J. L. 1966. Mechanisms of inheritance, 4. Extranuclear inheritance. In: *The fungi* **2**. Ed. by Ainsworth, G. C. & Sussman, A. S. 619–660. Academic Press, New York and London.

238 JOHNSTONE, K. I., TOWNSHEND, M. & WHITE, D. 1961. Inter-species change in thiobacilli. *J. gen. Microbiol.* **24**, 201–206.

239 JOHNSON, H. W., MEANS, U. M. & WEBER, C. R. 1965. Competition for nodule sites between strains of *Rhizobium japonicum* applied as inoculum and strains in the soil. *Agron. J.* **57**, 179–185.

240 JONES, D. & FARMER, V. C. 1967. The ecology and physiology of soil fungi involved in the degradation of lignin and related aromatic compounds. *J. Soil Sci.* **18**, 74–84.

241 JONES, D. & GRIFFITHS, E. 1964. The use of thin soil sections for the study of soil micro-organisms. *Pl. Soil* **20**, 232–240.

242 JONES, P. C. T. & MOLLISON, J. E. 1948. A technique for the quantitative estimation of soil micro-organisms. *J. gen. Microbiol.* **2**, 54–69.

243 KALELA, E. K. 1955. Die Veranderungen in den Wurzelverhaltnissen der Kiefernbestande im Laufe der Vegetationperiode. *Acta For. Fennica* **65**, 1–41.

244 KANAI, R., MIYACHI, S. & TAKAMIYA, A. 1960. Knall-gas reaction-linked fixation of labelled carbon dioxide in an autotrophic *Streptomyces. Nature, Lond.* **188**, 873–875.

245 KATZNELSON, H. & COLE, S. E. 1965. Production of gibberellin-like substances by bacteria and actinomycetes. *Can. J. Microbiol.* **11**, 733–741.

246 KATZNELSON, H. & RICHARDSON, L. T. 1948. Rhizosphere studies and associated microbiological phenomena in relation to strawberry root rot. *Sci. Agric.* **28**, 293–308.

247 KATZNELSON, H. & ROUATT, J. W. 1957. Manometric studies on rhizosphere and non-rhizosphere soil. *Can. J. Microbiol.* **3**, 673–678.

248 KEARNEY, P. C., KAUFMAN, D. D. & ALEXANDER, M. 1967. Biochemistry of herbicide decomposition in soils. In: *Soil Biochemistry.* Ed. by McLaren, A. D. & Peterson, G. H. 318–342. Arnold, London.

249 KENDRICK, W. B. 1959. The time factor in the decomposition of coniferous leaf litter. *Can. J. Bot.* **37**, 907–912.

250 KENDRICK, W. B. & BURGES, A. 1962. Biological aspects of decay of *Pinus sylvestris* leaf litter. *Nova Hedwig.* **4**, 313–342.

251 KENYAN, A., HENIS, Y. & KELLER, P. 1961. Factors influencing the composition of the cellulose-decomposing microflora on soil crumb plates. *Nature, Lond.* **191**, 307.

252 KITTAMS, H. A. & ATTOE, O. J. 1965. Availability of phosphorus in rock phosphate sulfur fusions. *Agron. J.* **57**, 331–334.

253 KO, W. H. & LOCKWOOD, J. L. 1967. Soil fungistasis; relation to fungal spore nutrition. *Phytopathology* **57**, 894–901.

254 KOBUS, J. 1962. The distribution of micro-organisms mobilizing phosphorus in different soils. *Acta. Microbiol. Pol.* **11**, 255–263.

255 KOWALSKI, M., HAM, G. E., FREDERICK, L. R. & ANDERSON, I. C. 1967. Relationship between strains of *Rhizobium japonicum* and their phages from soil and nodules of field-grown soybeans. *Bact. Proc.*, 4.

256 KRASILNIKOV, N. A. 1949. The role of micro-organisms in the weathering of rocks. *Mikrobiologiya* **18**, 318–323. (In Russian).

257 KROONTJE, W. & KEHR, W. R. 1956. Legume top and root yields

in the year of seeding and subsequent barley yield. *Agron. J.* **48**, 127–131.

258 KUNC, F. & MACURA, J. 1966. Decomposition of root exudates in soil. *Folia microbiol., Praha* **11**, 239–247.

259 KUYAMA, S. & PRAMER, D. 1962. Purification and properties of a protein having nemin activity. *Biochim. biophys. Acta* **56**, 631–632.

260 LAMPEN, J. O. 1965. Secretion of enzymes by micro-organisms. In: *Function and structure in micro-organisms.* Ed. by Pollock, M. R. & Richmond, M. H. *Symp. Soc. gen. Microbiol.* **15**, 115–133. Cambridge University Press, Cambridge.

261 LARSEN, S. 1967. Soil phosphorus. *Adv. Agron.* **19**, 151–210.

262 LEDEBERG, J. & LEDEBERG, E. M. 1952. Replica plating and indirect selection of bacterial mutants. *J. Bact.* **63**, 399–406.

263 LEES, H. & QUASTEL, J. H. 1946. Biochemistry of nitrification in soil. I. Kinetics of, and the effects of poisons on, soil nitrification as studied by a soil perfusion technique. *Biochem. J.* **40**, 803–815.

264 LENHARD, G. 1956. Die Dehydrogenaseaktivitat des Bodens als Mass für Microorganismentatigkeit in Boden. *Z. Pfl Ernähr. Düng. Bodenk.* **73**, 1–11.

265 LEWIS, D. H. & HARLEY, J. L. 1965a. Carbohydrate physiology of mycorrhizal roots of beech. I. Identity of endogenous sugars and utilization of exogenous sugars. *New. Phytol.* **64**, 224–237.

266 LEWIS, D. H. & HARLEY, J. L. 1965b. Carbohydrate physiology of mycorrhizal roots of beech. II. Utilization of exogenous sugars by uninfected and mycorrhizal roots. *New Phytol.* **64**, 238–255.

267 LEWIS, D. H. & HARLEY, J. L. 1965c. Carbohydrate physiology of mycorrhizal roots of beech. III. Movement of sugars between host and fungus. *New Phytol.* **64**, 256–269.

268 LEWIS, J. C., SNELL, N. S. & BURR, H. K. 1960. Water permeability of bacterial spores and the concept of a contractile cortex. *Science, NY.* **132**, 544–545.

269 LILLICH, T. & ELKAN, G. H. 1968. Role of polygalacturonase in invasion of root hairs of leguminous plants by *Rhizobium* sp. *Bact. Proc.*, 3.

270 LILLY, V. G. & LEONIAN, L. H. 1939. Vitamin B_1 in soil. *Science, NY.* **89**, 292.

271 LINFORD, M. B. 1937. Stimulated activity of natural enemies of nematodes. *Science, NY.* **85**, 123–124.

272 LINGAPPA, B. T. & LOCKWOOD, J. L. 1964. Activation of soil microflora by fungus spores in relation to soil fungistasis. *J. gen. Microbiol.* **35**, 215–228.

273 Lipman, J. G. & MacLean, H. 1916. The oxidation of sulfur in soils as a means of increasing the availability of mineral phosphates. *Soil Sci.* **1**, 533–540.

274 Lochhead, A. G. & Burton, M. O. 1955. Qualitative studies of soil micro-organisms. XII. Characteristics of vitamin B_{12} requiring bacteria. *Can. J. Microbiol.* **1**, 319–330.

275 Lochhead, A. G. 1957. Qualitative studies of soil microorganisms. XV. Capability of the predominant bacterial flora for synthesis of various growth factors. *Soil Sci.* **84**, 395–403.

276 Lochhead, A. G. & Thexton, R. H. 1947. Qualitative studies of soil micro-organisms. VII. The rhizosphere effect in relation to the amino acid nutrition of bacteria. *Can. J. Res.* **C25**, 20–26.

277 Lockwood, J. L. 1964. Soil fungistasis. *A. Rev. Phytopathol.* **2**, 341–362.

278 Lockwood, J. L. 1968. The fungal environment of soil bacteria. In: *The ecology of soil bacteria.* Ed. by Gray, T. R. G &. Parkinson, D. 44–65. Liverpool University Press, Liverpool.

279 Lockwood, J. L. & Lingappa, B. T. 1963. Fungitoxicity of sterilised soil inoculated with soil microflora. *Phytopathology* **53**, 917–920.

280 Loomis, W. E. 1949. Photosynthesis—an introduction. In: *Photosynthesis in plants.* Ed. by Franck, J. & Loomis, W. E. 1–17. Iowa State College Press, Ames.

281 Louw, H. A. & Webley, D. M. 1959a. The bacteriology of the root region of the oat plant grown under controlled pot culture conditions. *J. appl. Bact.* **22**, 216–226.

282 Louw, H. A. & Webley, D. M. 1959b. A study of soil bacteria dissolving certain phosphatic mineral fertilizers and related compounds. *J. appl. Bact.* **22**, 227–233.

283 Lynch, D. L. & Sears, O. H. 1952. The effect of inoculation upon yields of soybeans on treated and untreated soils. *Proc. Soil Sci. Soc. Am.* **16**, 124.

284 Lynch, D. L., Olney, H. O. & Wright, L. M. 1958. Some sugars and related carbohydrates found in Delaware soils. *J. Sci. Fd Agric.* **9**, 56–60.

285 Macauley, B. J. & Thrower, L. B. 1966. Succession of fungi in leaf litter of *Eucalyptus regnans.* *Trans. Br. mycol. Soc.* **49**, 509–520.

286 Macfadyen, A. 1963. The contribution of the microfauna to total soil metabolism. In: *Soil organisms.* Ed. by Doeksen, J. & van der Drift, J. 3–16. North Holland Publishing Co., Amsterdam.

287 Macfadyen, A. 1964. Energy flow in ecosystems and its exploitation

by grazing. In: *Grazing in terrestrial ecosystems.* Ed. by Crisp, D. J. 3–20. Blackwell, Oxford.

288 MACFADYEN, A. 1968. The animal habitat of soil bacteria. In: *The ecology of soil bacteria.* Ed. by Gray, T. R. G. & Parkinson, D. 66–76. Liverpool University Press, Liverpool.

289 MACFADYEN, A. 1970. Soil metabolism in relation to ecosystem energy flow and to primary and secondary production. In: *Methods for the study of production and energy flow in soil communities.* Ed. by Phillipson, J. Unesco, Paris.

290 MACURA, J. 1968. Physiological studies of rhizosphere bacteria. In: *The ecology of soil bacteria.* Ed. by Gray, T. R. G. & Parkinson, D. 379–395. Liverpool University Press, Liverpool.

291 MACURA, J., SZOLNOKI, J. & VANČURA, V. 1963. The decomposition of glucose in soil *Ust. Věd Inf. MZLVH* (*Rostl Výroba*) **36**, 788–792. (In Czech.)

292 MCCALLA, T. M., GUENZI, W. D. & NORSTADT, F. A. 1963. Microbial studies of phytotoxic substances in the stubble mulch system. *Z. allg. Mikrobiol.* **3**, 202–210.

293 MCCALLA, T. M. & HASKINS, F. A. 1961. Microorganisms and soil structure. *Miss. agric. Exp. Sta. Res. Bull.* **765**, 32–45.

294 MCCALLA, T. M. & HASKINS, F. A. 1964. Phytotoxic substances from soil micro-organisms and crop residues. *Bot. Rev.* **28**, 181–207.

295 MCCLUNG, N. M. 1960. Isolation of *Nocardia asteroides* from soils. *Mycologia* **52**, 154–156.

296 MCDOUGALL, B. M. & ROVIRA, A. D. 1965. Carbon-14 labelled photosynthate in wheat root exudates. *Nature, Lond.* **207**, 1104–1105.

297 MCFALL, E. & MANDELSTAM, J. 1963. Specific metabolic repression of three induced enzymes in *Escherichia coli. Biochem. J.* **89**, 391–398.

298 MCLAREN, A. D. 1960. Enzyme activity in structurally restricted systems. *Enzymologia* **21**, 356–364.

299 MCLAREN, A. D. 1968. Contribution to a discussion on the root region of plants. In: *The ecology of soil bacteria.* Ed. by Gray, T. R. G. & Parkinson, D. 434. Liverpool University Press, Liverpool.

300 MCLAREN, A. D., JENSEN, W. A. & JACOBSON, L. 1960. Absorption of enzymes and other proteins by barley roots. *Plant Physiol.* **35**, 549–556.

301 MCLAREN, A. D. & SKUJINS, J. J. 1963. Nitrification by *Nitrobacter agilis* on surfaces and in soil with respect to hydrogen ion concentration. *Can. J. Microbiol.* **9**, 729–731.

302 McLaren, A. D. & Skujins, J. 1968. The physical environment of micro-organisms in soil. In: *The ecology of soil bacteria.* Ed. by Gray, T. R. G. & Parkinson, D. 3–24. Liverpool University, Liverpool.

303 McLennan, E. 1928. The growth of fungi in the soil. *Ann. appl. Biol.* **15,** 95–109.

304 Mandelstam, J. 1969. Regulation of bacterial spore formation. In: *Microbial growth.* Ed. by Meadow, P. & Pirt, S. J. *Symp. Soc. gen. Microbiol.* **19,** 377–402. Cambridge University Press, Cambridge.

305 Mandelstam, J. & Halvorson, H. 1960. Turnover of protein and nucleic acid in soluble and ribosome fractions of non-growing *Escherichia coli. Biochim. biophys. Acta* **40,** 43–49.

306 Martin, J. P. 1950. Effects of fumigation and other soil treatments in the greenhouse on the fungus population of old citrus soil. *Soil Sci.* **69,** 107–122.

307 Martin, W. H. 1921. A comparison of inoculated and uninoculated sulfur for the control of potato scab. *Soil Sci.* **11,** 75–85.

308 Mathur, S. P. & Paul, E. A. 1967a. Microbial utilization of soil humic acids. *Can. J. Microbiol.* **13,** 573–580.

309 Mathur, S. P. & Paul, E. A. 1967b. Partial degradation of soil humic acids through biodegradation. *Can. J. Microbiol.* **13,** 581–586.

310 Mayaudon, J. & Simonart, P. 1958. Study of the decomposition of organic matter in soil by means of radioactive carbon. II. The decomposition of radioactive glucose in soil and distribution of radioactivity in the humus fractions of soil. *Pl. Soil* **9,** 376–380.

311 Mayaudon, J. & Simonart, P. 1959a. Étude de la décomposition de la matière organique dans le sol au moyen de carbone radioactif. IV. Décomposition des pigments fol. *Pl. Soil* **11,** 176–180.

312 Mayaudon, J. & Simonart, P. 1959b. Étude de la décomposition de la matière organique dans le sol au moyen de carbone radioactif. V. Décomposition de cellulose et de lignine. *Pl. Soil* **11,** 181–191.

313 Mayaudon, J. & Simonart, P. 1960. Decomposition of cellulose C^{14} and lignin C^{14} in the soil. In: *The ecology of soil fungi.* Ed. by Parkinson, D. & Waid, J. S. 257–259. Liverpool University Press, Liverpool.

314 Mazur, P. 1968. Survival of fungi after freezing and desiccation. In: *The fungi* **3.** Ed. by Ainsworth, G. S. & Sussman, A. S. 325–394. Academic Press, New York and London.

315 Means, U. M., Johnson, H. W. & Erdman, L. W. 1961. Competition between bacterial strains effecting nodulation in soybeans. *Proc. Soil Sci. Soc. Am.* **25**, 105–108.

316 Mechalas, R. J. & Rittenberg, S. C. 1960. Energy coupling in *Desulfovibrio desulfuricans*. *J. Bact.* **80**, 501–507.

317 Mehta, N. C., Streuli, H., Mueller, M. & Deuel, H. 1960. Role of polysaccharides in stabilization of natural soil aggregates. *J. Sci. Fd Agric.* **11**, 40–47.

318 Meiklejohn, J. 1957. Numbers of bacteria and actinomycetes in a Kenya soil. *J. Soil Sci.* **8**, 240–247.

319 Melin, E. 1954. Growth factor requirements of mycorrhizal fungi of forest trees. *Svensk bot. Tidskr.* **48**, 86–94.

320 Melin, E. & Das, V.S. R. 1954. The influence of root-metabolites on the growth of tree mycorrhizal fungi. *Physiol. Plant.* **7**, 851–858.

321 Meshkov, N. V. 1959. The effect of thiamin and biotin on the development of some soil microbes. *Mikrobiologiya* **28**, 894–899. (In Russian.)

322 Meynell, G. C. & Meynell, E. W. 1965. *Theory and practice in experimental bacteriology*. Cambridge University Press, Cambridge.

323 Miller, R. D. & Johnson, D. D. 1964. The effect of soil moisture tension on carbon dioxide evolution, nitrification and nitrogen mineralization. *Proc. Soil Sci. Soc. Am.* **28**, 644–647.

324 Morrill, L. G. & Dawson, J. E. 1962. Growth rates of nitrifying chemoautotrophs in soil. *J. Bact.* **83**, 205–206.

325 Moser, V. S. & Olson, R. V. 1953. Sulfur oxidation in four soils as influenced by soil moisture tension and sulfur bacteria. *Soil Sci.* **76**, 251–257.

326 Mosse, B. 1957. Growth and chemical composition of mycorrhizal and non-mycorrhizal apples. *Nature, Lond.* **179**, 922–924.

327 Müller, P. E. 1887. *Studiën über die natürlichen Humusformen.* Springer, Berlin.

328 Murphy, W. H. Jr. & Syverton, J. J. 1958. Absorption and translocation of mammalian viruses by plants. II. Recovery and distribution of viruses in plants. *Virology* **6**, 623–636.

329 Nicholas, D. J. D. 1965. Influence of the rhizosphere on the mineral nutrition of the plant. In: *Ecology of soil-borne plant pathogens*. Ed. by Baker, K. F. & Snyder, W. C. 210–217. University of California Press, Berkeley.

330 Nicholas, D. P., Parkinson, D. & Burges, N. A. 1965. Studies on fungi in a podzol. II. Application of the soil-sectioning technique to the study of amounts of fungal mycelium in the soil. *J. Soil Sci.* **16**, 258–269

331 NIELSEN, C. O. 1961. Respiratory metabolism of some populations of enchytraeid worms and free living nematodes. *Oikos* **12**, 17–35.

332 NIKITIN, D. I., VASIL'EVA, L. V. & LOKHMACHEVA, R. A. 1966. *New and rare forms of soil micro-organisms.* Academy of Sciences, USSR, Moscow. (In Russian).

333 NORMAN, A. G. 1943. The nitrogen nutrition of soybeans. I. Effect of inoculation and nitrogen fertilizer on the yield and composition of beans on Marshall silt loam. *Proc. Soil Sci. Soc. Am.* **8**, 226–228.

334 NORMAND, R. A. & COLMER, A. R. 1965. Oxidation of iron and sulfur in highly acidic soils of the Wilcox formation. *Bact. Proc.*, A23.

335 NOVAL, J. J. & NICKERSON, W. J. 1958. Decomposition of native keratin by *Streptomyces fradiae. J. Bact.* **27**, 251–263.

336 NUTMAN, P. S. 1956. The influence of the legume in root-nodule symbiosis. A comparative study of host determinants and functions. *Biol. Rev.* **31**, 109–151.

337 NUTMAN, P. S. 1958. The physiology of nodule formation. In: *Nutrition of the Legumes.* Ed. by E. G. Hallsworth. 87–107. Butterworth, London.

338 NUTMAN, P. S. 1965. The relation between nodule bacteria and the legume host in the rhizosphere and in the process of infection. In: *Ecology of soil-borne plant pathogens.* Ed. by Baker, K. F. & Snyder, W. C. 231–247. University of California Press, Berkeley.

339 NYE, P. H. 1968. Processes in the root environment. *J. Soil Sci.* **19**, 205–215.

340 O'CONNOR, F. B. 1963. Oxygen consumption and population metabolism of some populations of Enchytraeidae from North Wales. In: *Soil Organisms.* Ed. by Doeksen, J. & van der Drift, J. 32–48. North Holland Publishing Co., Amsterdam.

341 OKAFOR, N. 1966a. Ecology of micro-organisms on chitin buried in soil. *J. gen. Microbiol.* **44**, 311–327.

342 OKAFOR, N. 1966b. The ecology of micro-organisms on, and the decomposition of, insect wings in soil. *Pl. Soil* **25**, 211–237.

343 OLD, K. M. 1967. Effects of natural soil on survival of *Cochliobolus sativus. Trans. Br. mycol. Soc.* **50**, 615–624.

344 PANTOS, G. 1961. The vitamin-synthesizing capacity of some dominant strains of bacteria in the rhizosphere of wheat and maize. *Agrokhem. Talajt.* **10**, 511–522.

345 PARDEE, A. B. 1961. Response of enzyme synthesis and activity to environment. In: *Microbial reaction to environment.* Ed. by

Meynell, G. G. & Gooder, H. *Symp. Soc. gen. Microbiol.* **11**, 19–40. Cambridge University Press, Cambridge.

346 Park, D. 1955. Experimental studies on the ecology of fungi in soil. *Trans. Br. mycol. Soc.* **38**, 130–142.

347 Park, D. 1959. Some aspects of the biology of *Fusarium oxysporum* Schl. in soil. *Ann. Bot., Lond.* **NS23**, 35–49.

348 Park, D. 1960. Antagonism—the background to soil fungi. In: *The ecology of soil fungi.* Ed. by Parkinson, D. & Waid, J. S. 148–159. Liverpool University Press, Liverpool.

349 Park, D. 1965. Survival of microorganisms in soil. In: *Ecology of soil-borne plant pathogens.* Ed. by Baker, K. F. & Snyder, W. C. 82–97. University of California Press, Berkeley.

350 Parkinson, D. & Coups, E. 1963. Microbial activity in a podzol. In: *Soil organisms.* Ed. by Doeksen, J. and van der Drift, J. 167–175. North Holland Publishing Co., Amsterdam.

351 Parkinson, D. & Pearson, R. 1965. Factors affecting the stimulation of fungal development in the root region. *Nature, Lond.* **205**, 205–206.

352 Parkinson, D., Taylor, G. S. & Pearson, R. 1963. Studies on fungi in the root region. I. The development of fungi on young roots. *Pl. Soil* **19**, 332–349.

353 Parkinson, D. & Thomas, A. 1965. A comparison of methods for isolation of fungi from rhizospheres. *Can. J. Microbiol.* **11**, 1001–1008.

354 Parr, J. F. 1967. Biochemical considerations for increasing the efficiency of nitrogen fertilizers. *Soils Fertil., Harpenden* **30**, 207–213.

355 Parr, J. F., Parkinson, D. & Norman, A. G. 1967. Growth and activity of soil microorganisms in glass micro-beads. II. Oxygen uptake and direct observations. *Soil Sci.* **103**, 303–310.

356 Patrick, Z. A. 1955. The peach replant problem in Ontario. II. Toxic substances from microbial decomposition products of peach root residues. *Can. J. Bot.* **33**, 461–486.

357 Patrick, Z. A. & Toussoun, T. A. 1965. Plant residues and organic amendments in relation to biological control. In: *Ecology of soil-borne plant pathogens.* Ed. by Baker, K. F. & Snyder, W. C. 440–459. University of California Press, Berkeley.

358 Paul, E. A., Campbell, C. A., Rennie, D. A. & McCallum, K. J. 1964. Investigations of the dynamics of soil humus, utilizing carbon dating techniques. *Trans. 8th Int. Congr. Soil Sci., Bucharest* **3**, 201–208.

359 Pearson, R. & Parkinson, D. 1961. The sites of excretion of

ninhydrin-positive substances by broad bean seedlings. *Pl. Soil* **13**, 391–396.

360 Perfil'ev, B. V. & Gabe, D. R. 1961. *Capillary methods of studying micro-organisms.* Moscow, Leningrad: Academy of Sciences, USSR. (English translation, Oliver & Boyd, Edinburgh, University of Toronto Press.)

361 Person, C. 1968. Genetic adjustment of fungi to their environment. In: *The fungi* **3**. Ed. by Ainsworth, G. C. & Sussman, A. S. 395–415. Academic Press, New York and London.

362 Peterson, E. A. 1959. Seed-borne fungi in relation to colonization of roots. *Can. J. Microbiol.* **5**, 579–582.

363 Peterson, E. A. 1961. Observations on the influence of plant illumination on the fungal flora of roots. *Can. J. Microbiol.* **7**, 1–6.

364 Phillipson, J. 1966. *Ecological energetics.* Arnold, London.

365 Pinck, L. A. & Allison, F. E. 1944. The synthesis of lignin-like complexes by fungi. *Soil Sci.* **57**, 155–161.

366 Pirt, S. J. 1969. Microbial growth and product formation. In: *Microbial growth.* Ed. by Meadow, P. & Pirt, S. J. *Symp. Soc. gen. Microbiol.* **19**, 199–221. Cambridge University Press, Cambridge.

367 Pital, A., Janowitz, S. L., Hudak, C. E. & Lewis, E. E. 1966. Direct fluorescent labelling of microorganisms as a possible life detection technique. *Appl. Microbiol.* **14**, 119–123.

368 Pochon, J. 1957. Principe et applications d'une methodologie quantitative. *Pédologie* **7** (num. spéc.), 56–61.

369 Pochon, J. & Chalvignac, M. A. 1965. Biological oxidation of sulphur and degradation of vineyard soils. *Agrochimica* **9**, 155–161.

370 Pollock, M. R. 1967. Origin and function of penicillinase; a problem in biochemical evolution. *Br. med. J.* **4**, 71–77.

371 Polynov, B. B. 1945. The first stages of soil formation on massive crystalline rocks. *Pochvovedenie* (7), 327–339. (In Russian).

372 Pontecorvo, G. 1958. *Trends in genetic analysis.* Columbia University Press, New York.

373 Postgate, J. R. 1949. Competitive inhibition of sulphate reduction by selenate. *Nature, Lond.* **164**, 670–671.

374 Postgate, J. R. 1960. The economic activities of sulphate-reducing bacteria. *Progr. Ind. Microbiol.* **2**, 49–69.

375 Potgieter, H. J. & Alexander, M. 1966. Susceptibility and resistance of several fungi to microbial lysis. *J. Bact.* **91**, 1526–1532.

376 PRAMER, D. 1968. Significant syntheses by heterotrophic soil bacteria. In: *The ecology of soil bacteria.* Ed. Gray, T. R. G. & Parkinson, D. 220–233. Liverpool University Press, Liverpool.

377 PRAMER, D. & STARKEY, R. L. 1962. Determination of streptomycin in soil and the effect of soil colloidal material in its activity. *Soil Sci.* **94**, 48–54.

378 PRAMER, D. & STOLL, N. R. 1959. Nemin; a morphogenic substance causing trap formation by predaceous fungi. *Science, NY.* **129**, 966–967.

379 PRÉVOT, A. R. & POCHON, J. 1951. Étude de l'amylolyse bactérienne dans le sol par la méthode de la colonne de terre amidonée. *Annls Inst. Pasteur, Paris* **80**, 672–674.

380 PRINGSHEIM, E. G. 1946. *Pure cultures of algae: their preparation and maintenance.* Cambridge University Press, Cambridge.

381 PRINGSHEIM, E. G. 1949. Iron bacteria. *Biol. Rev.* **24**, 200–245.

382 PUGH, G. J. F. & MATHISON, G. E. 1962. Studies on fungi in coastal soils. III. An ecological survey of keratinophilic fungi. *Trans. Br. mycol. Soc.* **45**, 567–572.

383 PUTNAM, H. D. & SCHMIDT, E. L. 1959. Studies on the free amino acid fraction of soils. *Soil Sci.* **87**, 22–27.

384 QUASTEL, J. H. 1946. *Soil metabolism. Proc. Roy. Soc. Lond.* **B143**, 159–178.

385 RAHN, O. 1913. Die Bakterientatigkeit in Boden als Funktion der Nährungskonzentration und der unloslichen organischen Substanz. *Zentbl. Bakt. ParasitKde., Abt.* II **38**, 484–494.

386 RANEY, W. A. 1965. Physical factors of the soil as they affect micro-organisms. In: *Ecology of soil-borne plant pathogens.* Ed. by Baker, K. F. & Snyder, W. C. 115–118. University of California Press, Berkeley.

387 RANGASWAMI, G. & ETHIRAJ, S. 1962. Antibiotic production by *Streptomyces* sp. in unamended soil. *Phytopathology,* **52**, 989–992.

388 RAPER, J. R. 1966. Life cycles, basic patterns of sexuality, and sexual mechanisms. In *The fungi* **2**. Ed. by Ainsworth, G. C. & Sussman, A. S. 474–511. Academic Press, New York and London.

389 RASNIZINA, E. A. 1938. Formation of growth substances (auxin type) by bacteria. *C. r. (Dokl) Acad. Sci., URSS* **18**, 353–355.

390 RAW, F. 1962. Studies of earthworm populations. I. Leaf burial in apple orchards. *Ann. appl. Biol.* **50**, 389–404.

391 READ, M. P. 1953. The establishment of serologically identifiable strains of *Rhizobium trifolii* in field soils in competition with the native microflora. *J. gen. Microbiol.* **9**, 1–14.

392 ŘEHÁČEK, Z. 1956. Spreading of *Actinomyces* in the rhizosphere of some grains. *Cslka Microbiol.* **1**, 211–215.

393 REHM, H. 1959. Untersuchungen über des Verhalten von *Aspergillus niger* und einem *Streptomyces albus*—Stamm in Mischkultur. II. Mitteilung: Die Wechselbeziehungen im Erdboden. *Zentbl. Bakt. ParasitKde., Abt. II* **112**, 235–263.

394 REYNOLDS, E. R. C. 1966. The percolation of rain water through soil demonstrated by fluorescent dyes. *J. Soil Sci.* **17**, 127–132.

395 RICE, E. L. 1965. Inhibition of nitrogen-fixing and nitrifying bacteria by seed plants. II. Characterization and identification of inhibitors. *Physiol. Plant.* **18**, 255–268.

396 RIGHELATO, R. C., TRINCI, A. P. J., PIRT, S. J. & PEAT, A. 1968. The influence of maintenance energy and growth rate on the metabolic activities, morphology and conidiation of *Penicillium chrysogenum*. *J. gen. Microbiol.* **50**, 399–412.

397 ROBERTS, J. L. & ROBERTS, E. 1939. Auxin production by soil micro-organisms. *Soil Sci.* **48**, 135–139.

398 ROBINSON, J. D. 1963. Nitrification in a New Zealand grassland soil. *Pl. Soil* **19**, 173–183.

399 RODIN, L. E. & BAZILEVICH, N. I. 1967. *Production and mineral cycling in terrestrial vegetation.* Oliver & Boyd, Edinburgh.

400 ROPER, J. A. 1966. Mechanisms of inheritance, 3. The parasexual cycle. In: *The fungi* **2**. Ed. by Ainsworth, G. C. & Sussman, A. S. 589–617. Academic Press, New York and London.

401 ROSE, A. H. 1968. *Chemical microbiology* 2nd ed. Butterworth, London.

402 ROSE, R. E. 1957. Techniques for determining the effect of microorganisms on insoluble inorganic phosphates. *N.Z. J. Sci. Tech.* **B38**, 773–780.

403 ROSS, D. J. 1964. Effects of low-temperature storage on the oxygen uptake of soil. *Nature, Lond.* **204**, 503–504.

404 ROSS, D. J. 1965. A seasonal study of oxygen uptake of some pasture soils and activities of enzymes hydrolysing sucrose and starch. *J. Soil Sci.* **16**, 73–85.

405 ROSSI, J. & RICCARDO, S. 1927. Primi saggi di un metodo diretto per l'esame batteriologico de suolo. *Nuovi Ann. dell Agricolt.* **7**, 457–470.

406 ROUATT, J. W. & KATZNELSON, H. 1960. Influence of light on bacterial flora of roots. *Nature, Lond.* **186**, 659–660.

407 ROUATT, J. W. & KATZNELSON, H. 1961. A study of the bacteria on the root surface and in the rhizosphere of crop plants. *J. appl. Bact.* **24**, 164–171.

408 Rouatt, J. W., Katznelson, H. & Payne, T. M. B. 1960. Statistical evaluation of the rhizosphere effect. *Proc. Soil Sci. Soc. Am.* **24**, 271–273.

409 Roulet, M. A. 1948. Recherches sur les vitamines du sol. *Experientia*, **4**, 149–150.

410 Round, F. E. 1965. *The biology of the algae.* Arnold, London.

411 Rovira, A. D. 1956a. Plant root excretions in relation to the rhizosphere effect. I. The nature of root exudate from oats and peas. *Pl. Soil* **7**, 178–194.

412 Rovira, A. D. 1956b. Plant root excretions in relation to the rhizosphere effect. II. A study of the properties of root exudate and its effect on the growth of micro-organisms isolated from rhizosphere and control soil. *Pl. Soil* **7**, 195–208.

413 Rovira, A. D. 1956c. Plant root excretions in relation to the rhizosphere effect. III. The effect of root exudate on the numbers and activity of micro-organisms in soil. *Pl. Soil* **7**, 209–217.

414 Rovira, A. D. 1959. Root excretions in relation to the rhizosphere effect. IV. Influence of plant species, age of plant, temperature and calcium nutrition on exudation. *Pl. Soil* **11**, 53–64.

415 Rovira, A. D. 1962. Plant root exudates in relation to the rhizosphere microflora. *Soils Fertil., Harpenden* **25**, 167–172.

416 Rovira, A. D. 1965. Plant root exudates and their influence upon soil micro-organisms. In: *Ecology of soil-borne plant pathogens.* Ed. by Baker, K. F. & Snyder, W. C. 170–184. University of California Press, Berkeley.

417 Rovira, A. D. & Bowen, G. D. 1966. Phosphate incorporation by sterile and non-sterile plant roots. *Aust. J. biol. Sci.* **19**, 1167–1169.

418 Rovira, A. D. & Harris, J. R. 1961. Plant root excretions in relation to the rhizosphere effect. V. The exudation of B-group vitamins. *Pl. Soil* **14**, 199–214.

419 Rovira, A. D. & McDougall, B. M. 1967. Microbiological and biochemical aspects of the rhizosphere. In: *Soil Biochemistry.* Ed. by McLaren, A. D. & Peterson, G. H. 417–463. Arnold, London.

420 Rubenchik, L. I. 1963. *Azotobacter and its uses in agriculture.* Jerusalem: Israel program for scientific translations.

421 Ruinen, J. 1961. The phyllosphere. I. An ecologically neglected milieu. *Pl. Soil* **15**, 81–109.

422 Ruinen, J. 1963. Cuticle decomposition by microorganisms in the phyllosphere. *J. gen. Microbiol.* **32**, iv.

423 Russell, E. J. 1950. *Soil conditions and plant growth.* 8th ed. Longman, London.

424 RUSSELL, E. J. 1961. *Soil conditions and plant growth.* Rewritten by E. W. Russell. 9th ed. Longman, London.

425 RUSSELL, E. W. 1968. The agricultural environment of soil bacteria. In: *The ecology soil bacteria.* Ed. by Gray, T. R. G. & Parkinson, D. 77–89. Liverpool University Press, Liverpool.

426 SABININ, D. A. & MININA, E. G. 1932. Das mikrobiologische bodenprofil als zonales Kennzeichen. *Trans. 2nd Int. Congr. Soil Sci.* **3**, 224–235.

427 SADASIVAN, T. S. 1939. Succession of fungi decomposing wheat straw in different soils, with special reference to *Fusarium culmorum. Ann. appl. Biol.* **26**, 497–508.

428 SADASIVAN, T. S. 1965. Effect of mineral nutrients on soil micro-organisms and plant disease. In: *Ecology of soil-borne plant pathogens.* Ed. by Baker, K. F. & Snyder, R. C. 460–470. University of California Press, Berkeley.

429 SAHLMAN, H. & FÅHRAEUS, G. 1963. An electron microscope study of root-hair infection by *Rhizobium. J. gen. Microbiol.* **33**, 425–427.

430 SCHLEGEL, H. G. 1968. Significant chemolithotrophic reactions of bacteria in the soil. In: *The ecology of soil bacteria.* Ed. by Gray, T. R. G. & Parkinson, D. 234–255. Liverpool University Press, Liverpool.

431 SCHMIDT, E. L. 1950. Soil micro-organisms and plant growth substances. I. Historical. *Soil Sci.* **71**, 129–140.

432 SCHMIDT, E. L. 1954. Nitrate formation by a soil fungus. *Science, NY* **119**, 187–189.

433 SCHMIDT, E. L. & BANKOLE, R. O. 1963. The use of fluorescent antibody with the buried slide technique. In: *Soil organisms.* Ed. by Doeksen, J. & van der Drift, J. 197–204. North Holland Publishing Co., Amsterdam.

434 SCHMIDT, E. L., BANKOLE, R. & BOHLOOL, B. 1968. Fluorescent-antibody approach to study of rhizobia in soil. *J. Bact.* **95**, 1987–1992.

435 SCHMIDT, E. L., PUTNAM, H. D. & PAUL, E. A. 1960. Behaviour of free amino acids in soil. *Proc. Soil Sci. Soc. Am.* **24**, 107–109.

436 SCHMIDT, E. L. & STARKEY, R. L. 1951. Soil microorganisms and plant growth substances. II. Transformation of certain B-vitamins in soil. *Soil Sci.* **71**, 221–231.

437 SCHOTTEN, H. L. & STOKES, J. L. 1962. Isolation and properties of *Beggiatoa. Arch. Mikrobiol.* **42**, 353–368.

438 SEIFERT, J. 1961. The effect of low temperature on the intensity of nitrification. *Folia. Microbiol., Praha* **6**, 350–353.

439 SEIFERT, J. 1962. The influence of soil structure and moisture content on the number of bacteria and the degree of nitrification. *Folia Microbiol., Praha* **7**, 234–236.

440 SEQUEIRA, L. 1962. Influence of organic amendments on survival of *Fusarium oxysporum* f. *cubense* in the soil. *Phytopathology* **52**, 976–982.

441 SEWELL, G. W. F. & BROWN, J. C. 1960. Ecology of *Mucor ramannianus* Moeller. *Nature, Lond.* **183**, 1344–1345.

442 SHAMEEMULLAH, M. 1966. *The influence of soil moisture tension on the microbial populations of a pinewood soil.* Ph.D. thesis, University of Liverpool.

443 SHATTUCK, G. E. JR. & ALEXANDER, M. 1963. A differential inhibitor of nitrifying organisms. *Proc. Soil Sci. Soc. Am.* **27**, 600–601.

444 SHIELDS, L. M. & DURRELL, L. W. 1964. Algae in relation to soil fertility. *Bot. Rev.* **30**, 93–128.

445 SHILO, M. 1966. Predatory bacteria. *Science J.* **2** (9), 33–37.

446 SHTINA, E. A. 1954. The effect of agricultural plants on the algal flora of soil. *Trudy Kirov. s-kh. Inst.* **10**, 59–69.

447 SHUKLA, A. C. & GUPTA, A. B. 1967. Influence of algal growth promoting substances on growth, yield and protein contents of rice. *Nature, Lond.* **213**, 744.

448 SIMMONDS, P. M. 1947. The influence of antibiosis in the pathogenicity of *Helminthosporium sativum*. *Sci. Agric.* **27**, 625–632.

449 SKINNER, F. A. 1951. A method for distinguishing between viable spores and mycelial fragments of actinomycetes in soil. *J. gen. Microbiol.* **5**, 159–166.

450 SKINNER, F. A. 1968. The anaerobic bacteria of soil. In: *The ecology of soil bacteria.* Ed. by Gray, T. R. G. & Parkinson, D. 573–592. Liverpool University Press, Liverpool.

451 SKINNER, F. A., JONES, P. C. T. & MOLLISON, J. E. 1952. A comparison of a direct- and a plate-counting technique for quantitative estimation of soil micro-organisms. *J. gen. Microbiol.* **6**, 261–271.

452 SKINNER, F. A. & WALKER, N. 1961. Growth of *Nitrosomonas europaea* in batch and continuous culture. *Arch. Mikrobiol.* **38**, 339–349.

453 SKUJINS, J. J. 1967. Enzymes in soil. In: *Soil biochemistry.* Ed. by McLaren, A. D. & Peterson, G. H. 371–414. Arnold, London.

454 SKUJINS, J. J., POTGIETER, H. J. & ALEXANDER, M. 1965. Dissolution of fungal cell walls by a streptomycete chitinase and β-(1-3) glucanase. *Arch. Bioch. Biophys.* **111**, 358–364.

455 SMALII, V. T. & BERSHOVA, O. I. 1957. Formation of heteroauxin in *Azotobacter* cultures. *Mikrobiologiya* **26**, 520–526. (In Russian).

456 SMITH, S. E. 1966. Physiology and ecology of orchid mycorrhizal fungi with reference to seedling nutrition. *New Phytol.* **65**, 488–499.

457 SNEATH, P. H. A. 1962. Longevity of micro-organisms. *Nature, Lond.* **195**, 643–646.

458 SØRENSEN, H. 1962. Decomposition of lignin by soil bacteria and complex formation between autoxidized lignin and organic nitrogen compounds. *J. gen. Microbiol.* **27**, 21–34.

459 SØRENSEN, H. 1963. Studies on the decomposition of C^{14} labelled barley straw in soil. *Soil Sci.* **95**, 45–51.

460 SORIANO, S. & WALKER, N. 1964. Nitrifying organisms. In: *Rep. Rothamst. exp. Sta.*, 1963, *Soil microbiology report* by Nutman, P. S. 71.

461 SORIANO, S. & WALKER, N. 1968. Isolation of ammonia-oxidizing autotrophic bacteria. *J. appl. Bact.* **31**, 493–497.

462 SOULIDES, D. A. 1964. Antibiotics in soil. VI. Determination of micro-quantities of antibiotics in soil. *Soil Sci.* **97**, 286–289.

463 SOULIDES, D. A. 1965. Antibiotics in soil. VII. Production of streptomycin and tetracyclines in soil. *Soil Sci.* **100**, 200–206.

464 SOWDEN, F. J. 1955. Estimation of amino acids in soil hydrolysates by the Moore and Stein method. *Soil Sci.* **80**, 181–188.

465 SPERBER, J. I. 1958. Release of phosphate from soil minerals by hydrogen sulphide. *Nature, Lond.* **181**, 934.

466 STANIER, R. Y. 1942. The *Cytophaga* group: a contribution to the biology of myxobacteria. *Bact. Rev.* **6**, 143–196.

467 STANIER, R. Y., DOUDOROFF, M. & ADELBERG, E. A. 1963. *General microbiology*. 2nd ed. Macmillan, London.

468 STARKEY, R. L. 1929a. Some influences of the development of higher plants upon the micro-organisms in the soil. I. Historical and introductory. *Soil Sci.* **27**, 319–334.

469 STARKEY, R. L. 1929b. Some influences of the development of higher plants upon the micro-organisms in the soil. II. Influence of the stage of plant growth upon abundance of organisms. *Soil Sci.* **27**, 355–378.

470 STARKEY, R. L. 1929c. Some influences of the development of higher plants upon the micro-organisms in the soil. III. Influence of the stage of plant growth upon some activities of the organisms. *Soil Sci.* **27**, 433–444.

471 STARKEY, R. L. 1931. Some influences of the development of higher plants upon the micro-organisms in the soil. IV.

Influences of proximity of roots on abundance and activity of micro-organisms. *Soil Sci.* **32**, 367–393.

472 Starkey, R. L. 1938. Some influences of the development of higher plants upon the micro-organisms in the soil. VI. Microscopic examination of the rhizosphere. *Soil Sci.* **45**, 207–248.

473 Starkey, R. L. 1950. Relations of micro-organisms to transformations of sulfur in soil. *Soil Sci.* **70**, 55–65.

474 Starkey, R. L. 1958. The general physiology of the sulfate-reducing bacteria in relation to corrosion. *Prod. Monthly* **22**, (9), 12–30.

475 Starkey, R. L. 1968. The ecology of soil bacteria: discussion and concluding remarks. In: *The ecology of soil bacteria.* Ed. by Gray, T. R. G. & Parkinson, D. 635–646. Liverpool University Press, Liverpool.

476 Steinberg, R. A. 1952. Frenching symptoms produced in *Nicotiana tabacum* and *Nicotiana rustica* with optical isomers of isoleucine and leucine with *Bacillus cereus* toxin. *Pl. Physiol.* **27**, 302–308.

477 Stevenson, F. J. 1954. Ion exchange chromatography of the amino acids in soil hydrolysates. *Proc. Soil Sci. Soc. Am.* **18**, 373–376.

478 Stevenson, I. L. 1956. Antibiotic activity of actinomycetes in soil and their controlling effects on root-rot of wheat. *J. gen. Microbiol.* **14**, 440–448.

479 Stevenson, I. L. 1959. Dehydrogenase activity in soils. *Can. J. Microbiol.* **5**, 229–235.

480 Stevenson, I. L. & Ivarson, K. C. 1964. The decomposition of radioactive acetate in soils. *Can. J. Microbiol.* **10**, 139–142.

481 Steward, F. C. 1963. *Plant physiology* **3**. *Inorganic nutrition of plants.* Academic Press, New York and London.

482 Stewart, W. D. P. 1966. *Nitrogen fixation in plants.* Athlone Press, London.

483 Stewart, W. D. P. 1967. Transfer of biologically fixed nitrogen in a sand dune slack region. *Nature, Lond.* **214**, 603–604.

484 Stiven, G. 1952. Production of antibiotic substances by the roots of a grass (*Trachypogon phimosus* (H.B.K.) Nees) and of *Pentanisia variabilis* (E. Mey.) Harv. (Rubiaceae). *Nature, Lond.* **170**, 712–713.

485 Stolp, H. & Starr, M. P. 1963. *Bdellovibrio bacteriovorus* gen. et sp. n., a predatory, ectoparasitic and bacteriolytic micro-organism. *Antonie van Leeuwenhoek* **29**, 217–248.

486 Stotzky, G. 1965a. Replica plating techniques for studying microbial interactions in soil. *Can. J. Microbiol.* **11**, 629–636.

487 STOTZKY, G. 1965b. Microbial respiration. In: *Methods of soil analysis. II. Chemical and microbiological properties.* Ed. by Black, C. A. 1550–1570. American Society of Agronomy, Madison.

488 STOTZKY, G. 1966. Influence of clay minerals on micro-organisms. III. Effect of particle size, cation exchange capacity and surface area on bacteria. *Can. J. Microbiol.* **12**, 1235–1246.

489 STOTZKY, G. & GOOS, R. D. 1965. Effect of high CO_2 and low O_2 tensions on the soil microbiota. *Can. J. Microbiol.* **11**, 853–868.

490 STOTZKY, G., GOOS, R. D. & TIMONIN, M. J. 1962. Microbial changes in soil as a result of storage. *Pl. Soil* **16**, 1–18.

491 STOTZKY, G. & NORMAN, A. G. 1961. Factors limiting microbial activity in soil. II. The effect of sulfur. *Arch. Mikrobiol.* **40**, 370–382.

492 STOTZKY, G. & REM, L. T. 1966a. Effect of clay minerals on fungal respiration. *Bact. Proc.* 2–3.

493 STOTZKY, G. & REM, L. T. 1966b. Influence of clay minerals on micro-organisms. I. Montmorillonite and kaolinite on bacteria. *Can. J. Microbiol.* **12**, 547–564.

494 STOTZKY, G. & REM, L. T. 1967. Influence of clay minerals on micro-organisms. IV. Montmorillonite and kaolinite on fungi. *Can. J. Microbiol.* **13**, 1535–1550.

495 STRUGGER, S. 1948. Fluorescence microscope examination of bacteria in soil. *Can. J. Res.* **B26**, 188–193.

496 STÜVEN, K. 1960. Beiträge zur Physiologie und Systematik sulfatreduzierender Bakterien. *Arch. Mikrobiol.* **35**, 152–180.

497 SUSSMAN, A. S. 1968. Longevity and survivability of fungi. In: *The fungi* **3**. Ed. by Ainsworth, G. S. & Sussman, A. S. 447–486. Academic Press, New York and London.

498 SUSSMAN, A. S. & HALVORSON, H. O. 1966. *Spores, their dormancy and germination.* Harper & Row, New York and London.

499 SWABY, R. J. & LADD, J. N. 1962. Chemical nature, microbial resistance, and origin of soil humus. *Int. Soil Conf. N.Z.* 197–202. CSIRO, Adelaide.

500 SWABY, R. J. & PASSEY, B. I. 1953. A simple macrorespirometer for studies in soil microbiology. *Aust. J. agric. Res.* **4**, 334–339.

501 SZABÓ, I., MARTON, M. & PÁRTAI, G. 1964. Micro-milieu studies in the A-horizon of a mull-like rendzina. In: *Soil micromorphology.* Ed. by Jongerius, A. 33–45. Elsevier Publishing Co., Amsterdam.

502 SZABÓ, I., MARTON, M. & VARGA, L. 1964. Untersuchungen über die Hitzeresistenz, Temperatur- und Feuchtigkeitan-

spruche der Mikroorganismen eines mullartigen Waldrensina-bodens. *Pedobiologia* **4**, 43–66.

503 Takijima, Y. 1964. Studies on organic acids in paddy field soils with reference to their inhibitory effects on the growth of rice plants. *Soil Sci. Plant Nutr.* **10** (5), 14–29.

504 Taylor, G. S. & Parkinson, D. 1965. Studies on fungi in the root region. IV. Fungi associated with the roots of *Phaseolus vulgaris* L. *Pl. Soil* **22**, 1–20.

505 Teakle, D. S. 1962. Transmission of tobacco necrosis virus by a fungus, *Olpidium brassicae*. *Virology* **18**, 224–231.

506 Thomas, A., Nicholas, D. P. & Parkinson, D. 1965. Modifications of the agar film technique for assaying lengths of mycelium in soil. *Nature, Lond.* **205**, 105.

507 Thornton, H. G. & Gray, P. H. H. 1934. The numbers of bacterial cells in field soils, as estimated by the ratio method. *Proc. Roy. Soc.* **B115**, 522–543.

508 Thornton, R. H. 1952. The screened immersion plate. A method for isolating soil micro-organisms. *Research, Lond.* **5**, 190–191.

509 Timar, M. & Szabolics, I. 1964. The effect of organic matter on sulphate reduction in alkali (Szik) soils. *Agrokém. Talajt.* **13**, 129–136.

510 Timonin, M. I. 1941. The interaction of higher plants and soil microorganisms. III. Effect of by-products of plant growth on activity of fungi and actinomycetes. *Soil Sci.* **52**, 395–413.

511 Timonin, M. I. & Lochhead, A. G. 1948. Distribution of micro-organisms in the rhizosphere of a root system. *Trans. Roy. Soc. Can.* **42**, *Sect.* V, 175–181.

512 Tobback, P. & Laudelout, H. 1966. La culture de *Nitrobacter* en dialyse continue. *Arch. Mikrobiol.* **54**, 14–20.

513 Tribe, H. T. 1957. Ecology of micro-organisms in soils as observed during their development upon buried cellulose film. In: *Microbial ecology*. Ed. by Spicer, C. C. & Williams, R. E. O. *Symp. Soc. gen. Microbiol.* **7**, 287–298. Cambridge University Press, Cambridge.

514 Tribe, H. T. 1960a. Aspects of decomposition of cellulose in Canadian soils. I. Observations with the microscope. *Can. J. Microbiol.* **6**, 309–316.

515 Tribe, H. T. 1960b. Aspects of cellulose decomposition in Canadian soils. II. Nitrate nitrogen levels and carbon dioxide evolution. *Can. J. Microbiol.* **6**, 317–323.

516 Tribe, H. T. 1964. Microbial equilibrium in soil in relation to soil fertility. *Annls Inst. Pasteur, Paris* **107**, 698–710.

517 TRIBE, H. T. 1966. Interactions of soil fungi on cellulose film. *Trans. Br. mycol. Soc.* **49**, 457–466.

518 VÁGNEROVÁ, K. 1965. Properties of seed and soil bacteria with reference to the colonization of roots by micro-organisms. In: *Plant microbes relationships.* Ed. by Macura, J. & Vančura, V. 34–41. Czech Academy of Science, Prague.

519 VÁGNEROVÁ, K., MACURA, J. & ČATSKÁ, V. 1960a. Rhizosphere microflora of wheat. I. Composition of bacterial flora during the first stage of growth. *Folia microbiol., Praha* **5**, 298–310.

520 VÁGNEROVÁ, K., MACURA, J. & ČATSKÁ, V. 1960b. Rhizosphere microflora of wheat. II. Composition and properties of bacterial flora during the vegetation period of wheat. *Folia microbiol., Praha* **5**, 311–319.

521 VANČURA, V. 1961. Detection of gibberellic acid in *Azotobacter* cultures. *Nature, Lond.* **192**, 88–89.

522 VANČURA, V. & MACURA, J. 1960. Indole derivatives in *Azotobacter* cultures. *Folia microbiol., Praha* **5**, 294–297.

523 VAN GOOL, A. & LAUDELOUT, H. 1966. Formate utilization by *Nitrobacter winogradskyii. Biochim. Biophys. Acta.* **127**, 295–301.

524 VAN SCHREVEN, D. A. & HARMSEN, G. W. 1968. Soil bacteria in relation to the development of polders in the region of the former Zuider Zee. In: *The ecology of soil bacteria.* Ed. by Gray, T. R. G. & Parkinson, D. 474–499. Liverpool University Press, Liverpool.

525 VELDKAMP, H. 1955. A study of the aerobic decomposition of chitin by microorganisms. *Med. Landbouw., Wageningen* **55**, 127–174.

526 VELDKAMP, H. 1968. Bacterial physiology. In: *The ecology of soil bacteria.* Ed. by Gray, T. R. G. & Parkinson, D. 201–219. Liverpool University Press, Liverpool.

527 VENKATESAN, R. 1962. *Studies on the actinomycete population of paddy soil.* Annamalai University: Thesis.

528 VERHOEVEN, W., KOSTER, A. L. & VAN NIEVELT, M. C. A. 1954. Studies on true dissimilatory nitrate reduction. III. *Micrococcus denitrificans* Beijerinck, a bacterium capable of using molecular hydrogen in denitrification. *Antonie van Leeuwenhoek* **20**, 273–284.

529 VINCENT, J. M. 1954. The root-nodule bacteria of pasture legumes. *Proc. Linn. Soc. N.S.W.* **79**, iv-xxxii.

530 VINCENT, J. M. & WATERS, L. M. 1953. The influence of the host on competition amongst clover root-nodule bacteria. *J. gen. Microbiol.* **9**, 357–370.

531 VRANÝ, J. 1963. Effect of foliar application of urea on the root microflora. *Folia microbiol., Praha* **8**, 351–355.

532 VRANÝ, J. 1965. Effect of foliar application on the rhizosphere microflora. In: *Plant microbes relationships*. Ed. by Macura, J. & Vančura, V. 84–90. Czech Academy of Sciences, Prague.

533 WAID, J. S. 1957. Distribution of fungi within the decomposing tissues of rye-grass roots. *Trans. Br. mycol. Soc.* **40**, 391–406.

534 WAKSMAN, S. A. 1916. Do fungi live and produce mycelium in the soil? *Science, NY.* **44**, 320–322.

535 WAKSMAN, S. A. & DIEHM, R. A. 1931. On the decomposition of hemicelluloses by microorganisms. III. Decomposition of various hemicelluloses by aerobic and anaerobic bacteria. *Soil Sci.* **32**, 119–139.

536 WAKSMAN, S. A. & HUTCHINGS, I. J. 1937. Associative effects of micro-organisms. III. Associative and antagonistic relationships in the decomposition of plant residues. *Soil Sci.* **43**, 77–92.

537 WAKSMAN, S. A. & MARTIN, J. P. 1939. The role of micro-organisms in the conservation of soil. *Science, NY.* **90**, 304–305.

538 WANG, T. S. C., YANG, T. K. & CHUANG, T. T. 1967. Soil phenolic acids as plant growth inhibitors. *Soil Sci.* **103**, 239–246.

539 WARCUP, J. H. 1950. The soil-plate method for isolation of fungi from soil. *Nature, Lond.* **166**, 117–118.

540 WARCUP, J. H. 1955. Isolation of fungi from hyphae present in soil. *Nature, Lond.* **175**, 953–954.

541 WARCUP, J. H. 1957. Studies on the occurrence and activity of fungi in a wheat field soil. *Trans. Br. mycol. Soc.* **40**, 237–262.

542 WARCUP, J. H. 1960. Methods for isolation and estimation of activity of fungi in soil. In: *The ecology of soil fungi*. Ed. by Parkinson, D. & Waid, J. S. 3–21. Liverpool University Press, Liverpool.

543 WARCUP, J. H. 1965. Growth and reproduction of soil micro-organisms in relation to substrate. In: *Ecology of soil-borne plant pathogens*. Ed. by Baker, K. F. & Snyder, W. C. 52–68. University of California Press, Berkeley.

544 WARCUP, J. H. & BAKER, K. F. 1963. Occurrence of dormant ascospores in soil. *Nature, Lond.* **197**, 1317–1318.

545 WARCUP, J. H. & TALBOT, P. H. B. 1962. Ecology and identity of mycelia isolated from soil. *Trans. Br. mycol. Soc.* **45**, 495–518.

546 WARCUP, J. H. & TALBOT, P. H. B. 1963. Ecology and identity of mycelia isolated from soil. II. *Trans. Br. mycol. Soc.* **46**, 465–472.

547 WARCUP, J. H. & TALBOT, P. H. B. 1965. Ecology and identity of mycelia from soil. III. *Trans. Br. mycol. Soc.* **48**, 249–259.

548 WATANABE, A. 1960. Collection and cultivation of nitrogen-fixing blue-green algae and their effect on the growth and crop yield of rice plants. In: *P. Kachroo. Proc. Symp. Algol.* (1959). 162–166. Indian Council of Agricultural Research, New Delhi.

549 WATANABE, A. 1961. *Stud. Tokugawa Inst., Tokyo* **9**, 162. Cited by Stewart (1966).

550 WATANABE, A. & KIYOHARA, T. 1963. Symbiotic blue-green algae of lichens, liverworts and cycads. II. *Studies on microalgae and photosynthetic bacteria.* Jap. Soc. Pl. Physiol. 186–196. Tokyo University Press, Tokyo.

551 WATERHOUSE, D. F. 1957. Digestion in insects. *A. Rev. Entomol.* **2**, 1–18.

552 WATSON, P. 1965. Further observations on *Calcarisporium arbuscula*. *Trans. Br. mycol. Soc.* **48**, 9–17.

553 WEBER, C. R. 1966. Nodulating and non-nodulating soybean (*Glycine max*) isolines. I. Agronomic and chemical attributes. *Agron. J.* **58**, 43–46.

554 WEBLEY, D. M., EASTWOOD, D. J. & GIMMINGHAM, C. H. 1952. Development of a soil microflora in relation to plant succession in sand dunes including rhizosphere flora associated with colonizing species. *J. Ecol.* **40**, 168–178.

555 WEBLEY, D. M., HENDERSON, M. E. K. & TAYLOR, I. F. 1963. The microbiology of rocks and weathered stones. *J. Soil Sci.* **14**, 102–112.

556 WEISS, L. 1963. The pH value at the surface of *Bacillus subtilis*. *J. gen. Microbiol.* **32**, 331–340.

557 WEST, E. S. & TODD, W. R. 1961. *Textbook of biochemistry.* 3rd ed. Macmillan, New York.

558 WHYTE, R. O., NILSSON-LEISSNER, G. & TRUMBLE, H. C. 1953. *Legumes in agriculture.* FAO Agric. studies, Rome.

559 WIGGLESWORTH, V. B. 1948. The insect cuticle. *Biol. Rev.* **23**, 408–451.

560 WIGGLESWORTH, V. B. 1965. *The principles of insect physiology.* 6th ed. Methuen, London.

561 WILKINS, W. H. & HARRIS, G. C. M. 1947. The ecology of larger fungi. V. An investigation into the influence of rainfall and temperature on seasonal production of fungi in a beech wood and a pine wood. *Ann. appl. Biol.* **33**, 179–188.

562 WILKINSON, J. F. 1959. The problem of energy-storage compounds in bacteria. *Expl. Cell Res. Suppl.* **7**, 111–130.

563 WILLIAMS, S. T. 1962. *The soil washing technique and its application to a study of fungi in a podsolised soil.* Ph.D. thesis, University of Liverpool.

564 Williams, S. T. & Davies, F. L. 1965. Use of selective isolation and enumeration of actinomycetes in soil. *J. gen. Microbiol.* **38**, 251–261.

565 Williams, S. T. & Parkinson, D. 1964. Studies of fungi in a podzol. I. Nature and fluctuations of the fungus flora of the mineral horizons. *J. Soil Sci.* **15**, 331–341.

566 Williams, S. T., Parkinson, D. & Burges, N. A. 1965. An examination of the soil washing technique by its application to several soils. *Pl. Soil* **22**, 167–186.

567 Wilson, J. K. 1927. The number of ammonia-oxidizing organisms in soils. *1st Int. Congr. Soil Sci.* **3**, 14–22. Washington.

568 Winogradsky, S. 1924. Sur la microflore autochthone de la terre arable. *C. r. Acad. Sci., Paris* **178**, 1236–1239.

569 Winter, A. G. & Willeke, L. 1951. Über die Aufnahme von Antibiotics durch höhere Pflanzen und ihre Stabilität in natürlichen Böden. *Naturwissenschaften* **38**, 457–458.

570 Witkamp, M. 1966. Decomposition of leaf litter in relation to environment, microflora and microbial respiration. *Ecology* **47**, 194–201.

571 Woldendorp, J. W. 1963. The influence of living plants on denitrification. *Meded. LandbHoogesch., Wageningen* **63**, 1–100.

572 Wolk, P. 1965. Heterocyst germination under defined conditions. *Nature, Lond.* **205**, 201–202.

573 Wollum, A. G., Youngberg, C. T. & Gilmour, C. M. 1966. Characterization of a *Streptomyces* sp. isolated from root nodules of *Ceanothus velutinus* Dougl. *Proc. Soil Sci. Soc. Am.* **30**, 463–467.

574 Wood, R. K. S. 1959. Pathogen factors in the physiology of disease—pectic enzymes. In: *Plant pathology, problems and progress 1908–1958*. Ed. by Holton, C. S. *et al.* 100–109. University of Wisconsin Press, Madison.

575 Wright, J. M. 1956. The production of antibiotics in soil. III. Production of gliotoxin in wheat straw buried in soil. *Ann. appl. Biol.* **44**, 461–466.

576 Zvyagintsev, D. G. 1962. Adsorption of micro-organisms by soil particles. *Soviet Soil Sci.*, 140–144.

577 Babiuk, L. A. & Paul, E. A. 1970. The use of fluoroscein isothiocyanate in the determination of the bacterial biomass of grassland soil. *Can. J. Microbiol.* **16**, 57–62.

578 Gray, T. R. G. & Williams, S. T. 1971. Microbial productivity in soil. In: *Microbial productivity*. Ed. by Hughes, D. E. and Rose, A. *Symp. Soc. gen. Microbiol.* **21**. Cambridge University Press, Cambridge (in press).

579 Martin, M. M. *et al.* 1970. *Science*, **169**, 16.

MICRO-ORGANISM INDEX

SUBJECT INDEX

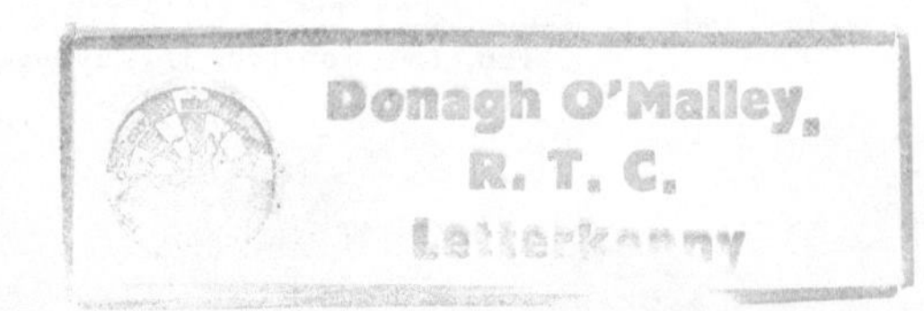